ESSAYS IN HONOR OF CARL G. HEMPEL

SYNTHESE LIBRARY

MONOGRAPHS ON EPISTEMOLOGY,

LOGIC, METHODOLOGY, PHILOSOPHY OF SCIENCE,

SOCIOLOGY OF SCIENCE AND OF KNOWLEDGE,

AND ON THE MATHEMATICAL METHODS OF

SOCIAL AND BEHAVIORAL SCIENCES

HUMANITIES PRESS / NEW YORK

ESSAYS IN HONOR OF CARL G. HEMPEL

A Tribute on the Occasion of his Sixty-Fifth Birthday

Essays by

DONALD DAVIDSON
ROBERT NOZICK
FREDERIC B. FITCH
PAUL OPPENHEIM
ADOLF GRÜNBAUM
HILARY PUTNAM
JAAKKO HINTIKKA
W. V. QUINE
RICHARD C. JEFFREY
NICHOLAS RESCHER
JAEGWON KIM
WESLEY C. SALMON
WILFRID SELLARS

Edited by Nicholas Rescher on behalf of the Editorial Committee:

ALAN ROSS ANDERSON
PAUL BENACERRAF
ADOLF GRÜNBAUM
GERALD J. MASSEY
NICHOLAS RESCHER
RICHARD S. RUDNER

D. REIDEL PUBLISHING COMPANY / DORDRECHT-HOLLAND

SOLE DISTRIBUTORS FOR U.S.A. AND CANADA
HUMANITIES PRESS / NEW YORK

Printed in The Netherlands by D. Reidel, Dordrecht

CARL G. HEMPEL

PREFACE

The eminent philosopher of science Carl G. Hempel, Stuart Professor of Philosophy at Princeton University and a Past President of the American Philosophical Association, has had a long and distinguished academic career in the course of which he has been professorial mentor to some of America's most distinguished philosophers. This volume gathers together twelve original papers by Hempel's students and associates into a volume intended to do homage to Hempel on the occasion of his 65th year in 1970. The papers are grouped around the unifying topic of Hempel's own interests in logic and philosophy of science, the great majority dealing with issues on inductive logic and the theory of scientific explanation – problems to which Hempel has devoted the bulk of his outstandingly fruitful efforts.

With the approach of 'Peter' Hempel's 65th birthday, an editorial committee sprang into being by an uncannily spontaneous process to prepare to commemorate this event with an appropriate *Festschrift*. The editors were pleased to receive unfailingly prompt and efficient cooperation on the part of all contributors. The responsibility of seeing the work through the press was assumed by Nicholas Rescher. The editors are grateful to all concerned for their collaboration.

ALAN ROSS ANDERSON
PAUL BENACERRAF
ADOLF GRÜNBAUM
GERALD J. MASSEY
NICHOLAS RESCHER
RICHARD S. RUDNER

TABLE OF CONTENTS

PAUL OPPENHEIM

REMINISCENCES OF PETER

The editors of this volume have asked me to contribute some personal reminiscence of 'Peter' Hempel (as he was and is called by all of us). It gives me great satisfaction to write such a note about Peter who has, for many years, been not only my admired and close friend, but also my philosophic associate and mentor. I hope that a chronology of our collaboration – with emphasis on *him* – will be of general interest, perhaps by shedding some new light on the professional development of this remarkable man.

We first met in Berlin, in 1933, at the suggestion of Hans Reichenbach, with whom I had worked for years on the possibility of a systematic ordering of the sciences. Peter agreed to continue these studies with me, especially with respect to the logic of ordering. As an opponent of the Nazi regime he was, at that time, prepared to leave Germany, where he had been headed for a secure teaching career. At the same time, I was forced to escape the Nazis and went to Brussels. This coincidence enabled me to prepare a refuge in Belgium for him, as well as an opportunity to continue his professional work through our collaboration. After several visits, he settled in Brussels in April 1934. There, for about three and a half years, we worked in fruitful collaboration, in which we were joined somewhat later by Kurt Grelling, who was particularly interested in Gestalt concepts.

The main results of our investigations have been published in two lectures[1] and in a book.[2] Our central topic was the logic and methodology of non-quantitative 'ordering concepts', or of 'comparative concepts', as Carnap has called them. We devoted special attention to the role that such concepts play in typological theories of human mental and physical constitution, such as those developed by Kretschmer and by Jaensch in Germany and later by Sheldon, Stevens, and Tucker in this country. Peter then applied the basic idea of ordering concepts to the theory of pluri-valued systems of propositional logic, by replacing the assumption that every sentence in the system has one of a finite or infinite set of

N. Rescher et al. (eds.), Essays in Honor of Carl G. Hempel.

truth-values by the weaker one that all the sentences under consideration are ordered by two relations, precedence (being less true) and coincidence (being equally true), so as to form a linear array according to increasing truth. On this weaker basis, it proved possible to construct analogues to the usual truth-tables and then to establish a set of logical truths for this system.[3]

The years in Brussels included also the very pleasant experience of joint visits to philosophical congresses in Paris, Copenhagen, and Cambridge; they offered us an opportunity to meet such scholars as Niels Bohr, J. B. S. Haldane, Jörgen Jörgensen, Otto Neurath, Karl R. Popper, and Susan Stebbing.

We also shared faithfully sad experiences. In this connection, I remember an evening when I was called to a hospital where Peter lay unconscious, breathing heavily, after having fractured his skull in a bicycle accident. The young intern in charge told me that nothing could be done for him, but I insisted that a senior physician be called immediately. Fortunately, by a daring procedure, he saved Peter's life.

In September 1937, while Grelling and I continued our work, Peter left for the United States, where he had been offered a year's research position at the University of Chicago. After his return to Brussels, we resumed our work until January 1939, when he moved definitively to the United States. There, he held teaching positions, first, at the City College of New York (1939–1940), and then at Queens College, New York (1940–1948). After that, he served on the faculties of Yale (1948–1955), and, from 1955 on, of Princeton. In August 1939, I moved to the United States as well and settled in Princeton. Here we were able to resume our close collaboration, which we had kept alive through correspondence.

One of the main problems still reflected our earlier interest in 'ordering concepts'. It concerned the development of a precise definition and a theory of a *quantitative concept of confirmation*. In this enterprise, we had the important help of Olaf Helmer who, at that time, was working with me in Princeton. We spent the early summer of 1944 with Olaf at Saranac Lake (where I introduced Peter to Einstein, who was then vacationing there). By mid-summer, we felt quite pleased to have formulated an explicit definition of 'degree of confirmation' for certain simple formalized languages. But, in the meantime, we heard from Carnap that he was at work on much the same problem, and we were curious – to say the least – whether he had been proceeding along the same lines. Fortunately, we were soon to have a chance to find out, for the Carnaps had invited Peter to spend the latter part of the summer with them in their vacation house at Santa Fé. We had agreed that Peter would wire Olaf and me 'Stop working' if it should turn out that Carnap had gotten far ahead of us or if he should find a decisive flaw in our approach. A few days later, the fateful wire did come. As Peter explained in a subsequent letter, Carnap had been amazed to find that we had indeed been thinking along very similar lines defining the concept of degree of confirmation as the quotient of two range measures; but he had pointed out that we had chosen a measure-function, which he attributed to Wittgenstein, and which incorporated, as he was able to show, a non-empiricist inductive policy, namely, to learn nothing from past experience.

In fact, however, we did not 'stop working', but changed our approach. Olaf provided the basic idea, which made use of R. A. Fisher's notion of maximum likelihood; this led to the definition and theoretical elaboration of a concept of degree of confirmation which did not have the objectionable non-empiricist feature, but which, in contrast to Carnap's concept of confirmation, did not have all the formal properties of a concept of probability. An account of our concept appeared in the same issue of *Philosophy of Science*[4] in which Carnap[5] first presented his approach to the problem. A more technical development of our concept of degree of confirmation was given in an article by Helmer and Oppenheim.[6] Carnap nicknamed our concept (because of the initials of the authors) 'the H_2O concept of confirmation'. We preferred not to speculate about a possible association with the German idiom 'in's Wasser gefallen', used to refer to an enterprise that has come to naught. At any rate, we often used the term in discussions among ourselves and with Carnap.

Peter and I quite naturally became more and more interested in the concept of *explanation* which, in some sense, is the converse of confirmation and had, as a matter of fact, already played a role in my work with Reichenbach on the ordering of sciences. The result of our discussions was published in *Philosophy of Science*.[7]

From then on, my close contact with Peter took another form, especially after he moved to Princeton. He very actively participated in the discussions of problems in many diverse fields which I explored with other collaborators.

In recounting my reminiscences of Peter, I have not emphasized our *personal* relationship. Despite the fact that I am much older than he is, I consider myself his pupil, especially in philosophy of science, and I should like to record here my great appreciation for all I have learned from such a highly intelligent, patient teacher. I am at least as grateful for the spiritual enrichment of my life that resulted from our friendship. In this, I had the opportunity to experience his extraordinary human qualities. Peter is tolerant; he listens with quick empathic understanding to people with opinions other than his own. He is absolutely reliable in both small and large matters, painstaking in his work, even at the cost of his leisure time. He is always prepared to go out of his way to help others as an aim in itself. To summarize his personal qualities, he is a fundamentally *good* man. In this connection, I am reminded of a discussion among Peter's friends about the question: Could Peter commit a murder? Two reasons for the negative answer were advanced: First, the evident one, his goodness, and second, jokingly, his lack of time. I wish him, for many years to come, the great happiness he deserves in his scholarly work as well as in his personal life.

Princeton, New Jersey

REFERENCES

[1] C. G. Hempel et Paul Oppenheim, 'L'importance logique de la notion de type', *Actes du Congrès International de Philosophie Scientifique*, vol. II, Hermann et Cie., Paris, 1935, pp. 1–9. Paul Oppenheim, 'Von Klassenbegriffen zu Ordnungsbegriffen', *Travaux du IXe Congrès International de Philosophie*, vol. VI, Hermann et Cie., Paris, 1937, pp. 69–76.

[2] Carl G. Hempel und Paul Oppenheim, *Der Typusbegriff im Lichte der neuen Logik*, A. W. Sijthoff, Leiden, 1936.

[3] Peter published his results in an article entitled 'A Purely Topological Form of Non-Aristotelian Logic', *The Journal of Symbolic Logic* **2** (1937), 97–112.

[4] Carl G. Hempel, and Paul Oppenheim, 'A Definition of "Degree of Confirmation"', *Philosophy of Science* **12** (1945), 98–115.

[5] Rudolf Carnap, 'On Inductive Logic', *Philosophy of Science* **12** (1945), 72–97.

[6] 'A Syntactical Definition of Probability and of Degree of Confirmation', *The Journal of Symbolic Logic* **10** (1945), 25–60.

[7] Carl G. Hempel and Paul Oppenheim, 'Studies in the Logic of Explanation', *Philosophy of Science* **15** (1948), 134–175.

W. V. QUINE

NATURAL KINDS

What tends to confirm an induction? This question has been aggravated on the one hand by Hempel's puzzle of the non-black non-ravens,[1] and exacerbated on the other by Goodman's puzzle of the grue emeralds.[2] I shall begin my remarks by relating the one puzzle to the other, and the other to an innate flair that we have for natural kinds. Then I shall devote the rest of the paper to reflections on the nature of this notion of natural kinds and its relation to science.

Hempel's puzzle is that just as each black raven tends to confirm the law that all ravens are black, so each green leaf, being a non-black non-raven, should tend to confirm the law that all non-black things are non-ravens, that is, again, that all ravens are black. What is paradoxical is that a green leaf should count toward the law that all ravens are black.

Goodman propounds his puzzle by requiring us to imagine that emeralds, having been identified by some criterion other than color, are now being examined one after another and all up to now are found to be green. Then he proposes to call anything *grue* that is examined today or earlier and found to be green or is not examined before tomorrow and is blue. Should we expect the first one examined tomorrow to be green, because all examined up to now were green? But all examined up to now were also grue; so why not expect the first one tomorrow to be grue, and therefore blue?

The predicate 'green', Goodman says,[3] is *projectible*; 'grue' is not. He says this by way of putting a name to the problem. His step toward solution is his doctrine of what he calls entrenchment,[4] which I shall touch on later. Meanwhile the terminological point is simply that projectible predicates are predicates ζ and η whose shared instances all do count, for whatever reason, toward confirmation of $\ulcorner$All ζ are $\eta\urcorner$.

Now I propose assimilating Hempel's puzzle to Goodman's by inferring from Hempel's that the complement of a projectible predicate need not be projectible. 'Raven' and 'black' are projectible; a black raven

N. Rescher et al. (eds.), Essays in Honor of Carl G. Hempel.

does count toward 'All ravens are black'. Hence a black raven counts also, indirectly, toward 'No non-black things are non-ravens', since this says the same thing. But a green leaf does not count toward 'All non-black things are non-ravens', nor, therefore, toward 'All ravens are black'; 'non-black' and 'non-raven' are not projectible. 'Green' and 'leaf' are projectible, and the green leaf counts toward 'All leaves are green' and 'All green things are leaves'; but only a black raven can confirm 'All ravens are black', the complements not being projectible.

If we see the matter in this way, we must guard against saying that a statement ⌜All ζ are η⌝ is law-like only if ζ and η are projectible. 'All non-black things are non-ravens' is a law despite its non-projectible terms, since it is equivalent to 'All ravens are black'. Any statement is law-like that is logically *equivalent* to ⌜All ζ are η⌝ for some projectible ζ and η.[5]

Having concluded that the complement of a projectible predicate need not be projectible, we may ask further whether there is *any* projectible predicate whose complement is projectible. I can conceive that there is not, when complements are taken strictly. We must not be misled by limited or relative complementation; 'male human' and 'non-male human' are indeed both projectible.

To get back now to the emeralds, why do we expect the next one to be green rather than grue? The intuitive answer lies in similarity, however subjective. Two green emeralds are more similar than two grue ones would be if only one were green. Green things, or at least green emeralds, are a kind.[6] A projectible predicate is one that is true of all and only the things of a kind. What makes Goodman's example a puzzle, however, is the dubious scientific standing of a general notion of similarity, or of kind.

The dubiousness of this notion is itself a remarkable fact. For surely there is nothing more basic to thought and language than our sense of similarity; our sorting of things into kinds. The usual general term, whether a common noun or a verb or an adjective, owes its generality to some resemblance among the things referred to. Indeed, learning to use a word depends on a double resemblance: first, a resemblance between the present circumstances and past circumstances in which the word was used, and second, a phonetic resemblance between the present utterance of the word and past utterances of it. And every reasonable expectation depends on resemblance of circumstances, together with our tendency to expect similar causes to have similar effects.

The notion of a kind and the notion of similarity or resemblance seem to be variants or adaptations of a single notion. Similarity is immediately definable in terms of kind; for things are similar when they are two of a kind. The very words for 'kind' and 'similar' tend to run in etymologically cognate pairs. Cognate with 'kind' we have 'akin' and 'kindred'. Cognate with 'like' we have 'ilk'. Cognate with 'similar' and 'same' and 'resemble' there are '*sammeln*' and 'assemble', suggesting a gathering into kinds.

We cannot easily imagine a more familiar or fundamental notion than this, or a notion more ubiquitous in its applications. On this score it is like the notions of logic: like identity, negation, alternation, and the rest. And yet, strangely, there is something logically repugnant about it. For we are baffled when we try to relate the general notion of similarity significantly to logical terms. One's first hasty suggestion might be to say that things are similar when they have all, or most, or many properties in common. Or, trying to be less vague, one might try defining comparative similarity – '*a* is more similar to *b* than to *c*' – as meaning that *a* shares more properties with *b* than with *c*. But any such course only reduces our problem to the unpromising task of settling what to count as a property.

The nature of the problem of what to count as a property can be seen by turning for a moment to set theory. Things are viewed as going together into sets in any and every combination, describable and indescribable. Any two things are joint members of any number of sets. Certainly then we cannot define '*a* is more similar to *b* than to *c*' to mean that *a* and *b* belong jointly to more sets than *a* and *c* do. If properties are to support this line of definition where sets do not, it must be because properties do not, like sets, take things in every random combination. It must be that properties are shared only by things that are significantly similar. But properties in such a sense are no clearer than kinds. To start with such a notion of property, and define similarity on that basis, is no better than accepting similarity as undefined.

The contrast between properties and sets which I suggested just now must not be confused with the more basic and familiar contrast between properties, as intensional, and sets as extensional. Properties are intensional in that they may be counted as distinct properties even though wholly coinciding in respect of the things that have them. There is no call to reckon kinds as intensional. Kinds can be seen as sets, determined by their members. It is just that not all sets are kinds.

If similarity is taken simple-mindedly as a yes-or-no affair, with no degrees, then there is no containing of kinds within broader kinds. For, as remarked, similarity now simply means belonging to some one same kind. If all colored things comprise a kind, then all colored things count as similar, and the set of all red things is too narrow to count as a kind. If on the other hand the set of all red things counts as a kind, then colored things do not all count as similar, and the set of all colored things is too broad to count as a kind. We cannot have it both ways. Kinds can, however, overlap; the red things can comprise one kind, the round another.

When we move up from the simple dyadic relation of similarity to the more serious and useful triadic relation of comparative similarity, a correlative change takes place in the notion of kind. Kinds come to admit now not only of overlapping but also of containment one in another. The set of all red things and the set of all colored things can now both count as kinds; for all colored things can now be counted as resembling one another more than some things do, even though less, on the whole, than red ones do.

At this point, of course, our trivial definition of similarity as sameness of kind breaks down; for almost any two things could count now as common members of some broad kind or other, and anyway we now want to define comparative or triadic similarity. A definition that suggests itself is this: *a* is more similar to *b* than to *c* when *a* and *b* belong jointly to more kinds than *a* and *c* do. But even this works only for finite systems of kinds.

The notion of kind and the notion of similarity seemed to be substantially one notion. We observed further that they resist reduction to less dubious notions, as of logic or set theory. That they at any rate be definable each in terms of the other seems little enough to ask. We just saw a somewhat limping definition of comparative similarity in terms of kinds. What now of the converse project, definition of kind in terms of similarity?

One may be tempted to picture a kind, suitable to a comparative similarity relation, as any set which is 'qualitatively spherical' in this sense: it takes in exactly the things that differ less than so-and-so much from some central norm. If without serious loss of accuracy we can assume that there are one or more actual things (*paradigm cases*) that nicely exemplify the desired norm, and one or more actual things (*foils*) that deviate just barely too much to be counted into the desired kind at all, then our definition is easy: *the kind with paradigm a and foil b* is the set of all the things to which *a* is more similar than *a* is to *b*. More generally, then, a set may be said to be a *kind* if and only if there are *a* and *b*, known or unknown, such that the set is the kind with paradigm *a* and foil *b*.

If we consider examples, however, we see that this definition does not give us what we want as kinds. Thus take red. Let us grant that a central shade of red can be picked as norm. The trouble is that the paradigm cases, objects in just that shade of red, can come in all sorts of shapes, weights, sizes, and smells. Mere degree of overall similarity to any one such paradigm case will afford little evidence of degree of redness, since it will depend also on shape, weight, and the rest. If our assumed relation of comparative similarity were just comparative chromatic similarity, then our paradigm-and-foil definition of kind would indeed accommodate redkind. What the definition will not do is distill purely chromatic kinds from mixed similarity.

A different attempt, adapted from Carnap, is this: a set is a kind if all its members are more similar to one another than they all are to any one thing outside the set. In other words, each non-member differs more from some member than that member differs from any member. However, as Goodman showed in a criticism of Carnap,[7] this construction succumbs to what Goodman calls the difficulty of imperfect community. Thus consider the set of all red round things, red wooden things, and round wooden things. Each member of this set resembles each other member somehow: at least in being red, or in being round, or in being wooden, and perhaps in two or all three of these respects or others. Conceivably, moreover, there is no one thing outside the set that resembles every member of the set to even the least of these degrees. The set then meets the proposed definition of kind. Yet surely it is not what anyone means by a kind. It admits yellow croquet balls and red rubber balls while excluding yellow rubber balls.

The relation between similarity and kind, then, is less clear and neat than could be wished. Definition of similarity in terms of kind is halting, and definition of kind in terms of similarity is unknown. Still the two notions are in an important sense correlative. They vary together. If we reassess something *a* as less similar to *b* than to *c*, where it had counted as more similar to *b* than to *c*, surely we will correspondingly permute *a*, *b*, and *c* in respect of their assignment to kinds; and conversely.

I have stressed how fundamental the notion of similarity or of kind is to our thinking, and how alien to logic and set theory. I want to go on now to say more about how fundamental these notions are to our thinking, and something also about their non-logical roots. Afterward I want to bring out how the notion of similarity or of kind changes as science progresses. I shall suggest that it is a mark of maturity of a branch of science that the notion of similarity or kind finally dissolves, so far as it is relevant to that branch of science. That is, it ultimately submits to analysis in the special terms of that branch of science and logic.

For deeper appreciation of how fundamental similarity is, let us observe more closely how it figures in the learning of language. One learns by *ostension* what presentations to call yellow; that is, one learns by hearing the word applied to samples. All he has to go on, of course, is the similarity of further cases to the samples. Similarity being a matter of degree, one has to learn by trial and error how reddish or brownish or greenish a thing can be and still be counted yellow. When he finds he has applied the word too far out, he can use the false cases as samples to the contrary; and then he can proceed to guess whether further cases are yellow or not by considering whether they are more similar to the in-group or the out-group. What one thus uses, even at this primitive stage of learning, is a fully functioning sense of similarity, and relative similarity at that: *a* is more similar to *b* than to *c*.

All these delicate comparisons and shrewd inferences about what to call yellow are, in Sherlock Holmes's terminology, elementary. Mostly the process is unconscious. It is the same process by which an animal learns to respond in distinctive ways to his master's commands or other discriminated stimulations.

The primitive sense of similarity that underlies such learning has, we saw, a certain complexity of structure: *a* is more similar to *b* than to *c*. Some people have thought that it has to be much more complex still:

that it depends irreducibly on *respects*, thus similarity in color, similarity in shape, and so on. According to this view, our learning of yellow by ostension would have depended on our first having been told or somehow apprised that it was going to be a question of color. Now hints of this kind are a great help, and in our learning we often do depend on them. Still one would like to be able to show that a single general standard of similarity, but of course comparative similarity, is all we need, and that respects can be abstracted afterward. For instance, suppose the child has learned of a yellow ball and block that they count as yellow, and of a red ball and block that they do not, and now he has to decide about a yellow cloth. Presumably he will find the cloth more similar to the yellow ball and to the yellow block than to the red ball or red block; and he will not have needed any prior schooling in colors and respects. Carnap undertook to show long ago how some respects, such as color, could by an ingenious construction be derived from a general similarity notion;[8] however, this development is challenged, again, by Goodman's difficulty of imperfect community.

A standard of similarity is in some sense innate. This point is not against empiricism; it is a commonplace of behavioral psychology. A response to a red circle, if it is rewarded, will be elicited again by a pink ellipse more readily than by a blue triangle; the red circle resembles the pink ellipse more than the blue triangle. Without some such prior spacing of qualities, we could never acquire a habit; all stimuli would be equally alike and equally different. These spacings of qualities, on the part of men and other animals, can be explored and mapped in the laboratory by experiments in conditioning and extinction.[9] Needed as they are for all learning, these distinctive spacings cannot themselves all be learned; some must be innate.

If then I say that there is an innate standard of similarity, I am making a condensed statement that can be interpreted, and truly interpreted, in behavioral terms. Moreover, in this behavioral sense it can be said equally of other animals that they have an innate standard of similarity too. It is part of our animal birthright. And, interestingly enough, it is characteristically animal in its lack of intellectual status. At any rate we noticed earlier how alien the notion is to mathematics and logic.

This innate qualitative spacing of stimulations was seen to have one of its human uses in the ostensive learning of words like 'yellow'. I should add as a cautionary remark that this is not the only way of learning words, nor the commonest; it is merely the most rudimentary way. It works when the question of the reference of a word is a simple question of spread: how much of our surroundings counts as yellow, how much counts as water, and so on. Learning a word like 'apple' or 'square' is more complicated, because here we have to learn also where to say that one apple or square leaves off and another begins. The complication is that apples do not add up to an apple, nor squares, generally, to a square. 'Yellow' and 'water' are mass terms, concerned only with spread; 'apple' and 'square' are terms of divided reference, concerned with both spread and individuation. Ostension figures in the learning of terms of this latter kind too, but the process is more complex.[10] And then there are all the other sorts of words, all those abstract and neutral connectives and adverbs and all the recondite terms of scientific theory; and there are also the grammatical constructions themselves to be mastered. The learning of these things is less direct and more complex still. There are deep problems in this domain, but they lie aside from the present topic.

Our way of learning 'yellow', then, gives less than a full picture of how we learn language. Yet more emphatically, it gives less than a full picture of the human use of an innate standard of similarity, or innate spacing of qualities. For, as remarked, every reasonable expectation depends on similarity. Again on this score, other animals are like man. Their expectations, if we choose so to conceptualize their avoidance movements and salivation and pressing of levers and the like, are clearly dependent on their appreciation of similarity. Or, to put matters in their methodological order, these avoidance movements and salivation and pressing of levers and the like are typical of what we have to go on in mapping the animals' appreciation of similarity, their spacing of qualities.

Induction itself is essentially only more of the same: animal expectation or habit formation. And the ostensive learning of words is an implicit case of induction. Implicitly the learner of 'yellow' is working inductively toward a general law of English verbal behavior, though a law that he will never try to state; he is working up to where he can in general judge when an English speaker would assent to 'yellow' and when not.

Not only is ostensive learning a case of induction; it is a curiously comfortable case of induction, a game of chance with loaded dice. At any rate this is so if, as seems plausible, each man's spacing of qualities is enough like his neighbor's. For the learner is generalizing on his yellow samples by similarity considerations, and his neighbors have themselves acquired the use of the word 'yellow', in their day, by the same similarity considerations. The learner of 'yellow' is thus making his induction in a friendly world. Always, induction expresses our hope that similar causes will have similar effects; but when the induction is the ostensive learning of a word, that pious hope blossoms into a foregone conclusion. The uniformity of people's quality spaces virtually assures that similar presentations will elicit similar verdicts.

It makes one wonder the more about other inductions, where what is sought is a generalization not about our neighbor's verbal behavior but about the harsh impersonal world. It is reasonable that our quality space should match our neighbor's, we being birds of a feather; and so the general trustworthiness of induction in the ostensive learning of words was a put-up job. To trust induction as a way of access to the truths of nature, on the other hand, is to suppose, more nearly, that our quality space matches that of the cosmos. The brute irrationality of our sense of similarity, its irrelevance to anything in logic and mathematics, offers little reason to expect that this sense is somehow in tune with the world – a world which, unlike language, we never made. Why induction should be trusted, apart from special cases such as the ostensive learning of words, is the perennial philosophical problem of induction.

One part of the problem of induction, the part that asks why there should be regularities in nature at all, can, I think, be dimisssed. *That* there are or have been regularities, for whatever reason, is an established fact of science; and we cannot ask better than that. *Why* there have been regularities is an obscure question, for it is hard to see what would count as an answer. What does make clear sense is this other part of the problem of induction: why does our innate subjective spacing of qualities accord so well with the functionally relevant groupings in nature as to make our inductions tend to come out right? Why should our subjective spacing of qualities have a special purchase on nature and a lien on the future?

There is some encouragement in Darwin. If people's innate spacing of qualities is a gene-linked trait, then the spacing that has made for the most successful inductions will have tended to predominate through natural selection.[11] Creatures inveterately wrong in their inductions have a pathetic but praiseworthy tendency to die before reproducing their kind.

At this point let me say that I shall not be impressed by protests that I am using inductive generalizations, Darwin's and others, to justify induction, and thus reasoning in a circle. The reason I shall not be impressed by this is that my position is a naturalistic one; I see philosophy not as an *a priori* propaedeutic or groundwork for science, but as continuous with science. I see philosophy and science as in the same boat – a boat which, to revert to Neurath's figure as I so often do, we can rebuild only at sea while staying afloat in it. There is no external vantage point, no first philosophy. All scientific findings, all scientific conjectures that are at present plausible, are therefore in my view as welcome for use in philosophy as elsewhere. For me, then, the problem of induction is a problem about the world: a problem of how we, as we now are (by our present scientific lights), in a world we never made, should stand better than random or coin-tossing chances of coming out right when we predict by inductions which are based on our innate, scientifically unjustified similarity standards. Darwin's natural selection is a plausible partial explanation.

It may, in view of a consideration to which I next turn, be almost explanation enough. This consideration is that induction, after all, has its conspicuous failures. Thus take color. Nothing in experience, surely, is more vivid and conspicuous than color and its contrasts. And the remarkable fact, which has impressed scientists and philosophers as far back at least as Galileo and Descartes, is that the distinctions that matter for basic physical theory are mostly independent of color contrasts. Color impresses man; raven black impresses Hempel; emerald green impresses Goodman. But color is cosmically secondary. Even slight differences in sensory mechanisms from species to species, Smart remarks,[12] can make overwhelming differences in the grouping of things by color. Color is king in our innate quality space, but undistinguished in cosmic circles. Cosmically, colors would not qualify as kinds.

Color is helpful at the food-gathering level. Here it behaves well under induction, and here, no doubt, has been the survival value of our color-slanted quality space. It is just that contrasts that are crucial for such activities can be insignificant for broader and more theoretical science. If man were to live by basic science alone, natural selection would shift its support to the color-blind mutation.

Living as he does by both bread and basic science, man is torn. Things about his innate similarity standards that are helpful in the one sphere can be a hindrance in the other. Credit is due to man's inveterate ingenuity, or human sapience, for having worked around the blinding dazzle of color vision and found the more significant regularities elsewhere. Evidently natural selection has dealt with the conflict by endowing man doubly: with both a color-slanted quality space and the ingenuity to rise above it.

He has risen above it by developing modified systems of kinds, hence modified similarity standards for scientific purposes. By the trial-and-error process of theorizing he has regrouped things into new kinds which prove to lend themselves to many inductions better than the old.

A crude example is the modification of the notion of fish by excluding whales and porpoises. Another taxonomic example is the grouping of kangaroos, opossums, and marsupial mice in a single kind, marsupials, while excluding ordinary mice. By primitive standards the marsupial mouse is more similar to the ordinary mouse than to the kangaroo; by theoretical standards the reverse is true.

A theoretical kind need not be a modification of an intuitive one. It may issue from theory full-blown, without antecedents; for instance the kind which comprises positively charged particles. We revise our standards of similarity or of natural kinds on the strength, as Goodman remarks,[13] of second-order inductions. New groupings, hypothetically adopted at the suggestion of a growing theory, prove favorable to inductions and so become 'entrenched'. We newly establish the projectibility of some predicate, to our satisfaction, by successfully trying to project it. In induction nothing succeeds like success.

Between an innate concept of similarity or spacing of qualities and a scientifically sophisticated one, there are all gradations. Science, after all, differs from common sense only in degree of methodological sophistication. Our experiences from earliest infancy are bound to have overlaid our innate spacing of qualities by modifying and supplementing our grouping habits little by little, inclining us more and more to an appreciation of theoretical kinds and similarities, long before we reach the point of studying science systematically as such. Moreover, the later phases do not wholly supersede the earlier; we retain different similarity standards, different systems of kinds, for use in different contexts. We all still say that a marsupial mouse is more like an ordinary mouse than a kangaroo, except when we are concerned with genetic matters. Something like our innate quality space continues to function alongside the more sophisticated regroupings that have been found by scientific experience to facilitate induction.

We have seen that a sense of similarity or of kinds is fundamental to learning in the widest sense – to language learning, to induction, to expectation, Toward a further appreciation of how utterly this notion permeates our thought, I want now to point out a number of other very familiar and central notions which seem to depend squarely on this one. They are notions that are definable in terms of similarity, or kinds, and further irreducible.

A notable domain of examples is the domain of dispositions, such as Carnap's example of solubility in water. To say of some individual object that it is soluble in water is not to say merely that it always dissolves when in water, because this would be true by default of any object, however insoluble, if it merely happened to be destined never to get into water. It is to say rather that it *would* dissolve if it were in water; but this account brings small comfort, since the device of a subjunctive conditional involves all the perplexities of disposition terms and more. Thus far I simply repeat Carnap.[14] But now I want to point out what could be done in this connection with the notion of kind. Intuitively, what qualifies a thing as soluble though it never gets into water is that it is of the same kind as the things that actually did or will dissolve; it is similar to them. Strictly we cannot simply say '*the* same kind', nor simply 'similar', when we have wider and narrower kinds, less and more similarity. Let us then mend our definition by saying that the soluble things are the common members of *all* such kinds. A thing is soluble if *each* kind that is broad enough to embrace all actual victims of solution embraces it too.

Graphically the idea is this: we make a set of all the sometime victims, all the things that actually did or will dissolve in water, and then we add just enough other things to round the set out into a kind. This is the water-soluble kind.

If this definition covers just the desired things, the things that are really soluble in water, it owes its success to a circumstance that could be otherwise. The needed circumstance is that a sufficient variety of things actually get dissolved in water to assure their not all falling under any one kind narrower than the desired water-soluble kind itself. But it is a plausible circumstance, and I am not sure that its accidental character is a drawback. If the trend of events had been otherwise, perhaps the solubility concept would not have been wanted.

However, if I seem to be defending this definition, I must now hasten to add that of course it has much the same fault as the definition which used the subjunctive conditional. This definition uses the unreduced notion of kind, which is certainly not a notion we want to rest with either; neither theoretical kind nor intuitive kind. My purpose in giving the definition is only to show the link between the problem of dispositions and the problem of kinds.

As between theoretical and intuitive kinds, certainly the theoretical ones are the ones wanted for purposes of defining solubility and other dispositions of scientific concern. Perhaps 'amiable' and 'reprehensible' are disposition terms whose definitions should draw rather on intuitive kinds.

In considering the disposition of solubility we observed a link first with the subjunctive conditional and then with the notion of kind. This suggests comparing also the two end terms, so as to see the connection between the subjunctive conditional and the notion of kind. We had then, on the one side, the subjunctive conditional 'If *x* were in water it would dissolve'; and on the other side, in terms of kinds, we had 'Each kind that embraces all things that ever get into water and dissolve, embraces *x*'. Here we have equated a sample subjunctive conditional to a sentence about kinds. We can easily enough generalize the equivalence to cover a significant class of subjunctive conditionals: the form 'If *x* were an *F*, then *x* would be a *G*' gets equated to 'Each kind that embraces all *F*'s that are *G*'s embraces *x*'. Notice that the *F*'s themselves, here, would not be expected to constitute a kind; nor the *G*'s; nor the *F*'s which are *G*'s. But you take the fewest things you can which, added to the *F*'s which are *G*'s, suffice to round the set out to a kind. Then *x* is one of these few additional things; this is the interpretation we get of the subjunctive conditional 'If *x* were an *F*, then *x* would be a *G*'.

One might try this formula out on other examples, and study it for possible light on subjunctive conditionals more generally. Some further insight into this queer idiom might thus be gained. But let us remember that we are still making uncritical use of the unreduced notion of kind. My purpose, again, is only to show the link between these matters.

Another dim notion, which has intimate connections with dispositions and subjunctive conditionals, is the notion of cause; and we shall see that it too turns on the notion of kinds. Hume explained cause as invariable succession, and this makes sense as long as the cause and effect are referred to by general terms. We can say that fire causes heat, and we can mean thereby, as Hume would have it, that each event classifiable under the head of fire is followed by an event classifiable under the head of heat, or heating up. But this account, whatever its virtues for these general causal statements, leaves singular causal statements unexplained.

What does it mean to say that the kicking over of a lamp in Mrs. Leary's barn caused the Chicago fire? It cannot mean merely that the event at Mrs. Leary's belongs to a set, and the Chicago fire belongs to a set, such that there is invariable succession between the two sets: every member of the one set is followed by a member of the other. This paraphrase is trivially true and too weak. Always, if one event happens to be followed by another, the two belong to *certain* sets between which there is invariable succession. We can rig the sets arbitrarily. Just put any arbitrary events in the first set, including the first of the two events we are interested in; and then in the other set put the second of those two events, together with other events that happen to have occurred just after the other members of the first set.

Because of this way of trivialization, a singular causal statement says no more than that the one event was followed by the other. That is, it says no more if we use the definition just now contemplated; which, therefore, we must not. The trouble with that definition is clear enough: it is the familiar old trouble of the promiscuity of sets. Here, as usual, kinds, being more discriminate, enable us to draw distinctions where sets do not. To say that one event caused another is to say that the two events are of *kinds* between which there is invariable succession. If this correction does not yet take care of Mrs. Leary's cow, the fault is only with invariable succession itself, as affording too simple a definition of general causal statements; we need to hedge it around with provisions for partial or contributing causes and a good deal else. That aspect of the causality problem is not my concern. What I wanted to bring out is just the relevance of the notion of kinds, as the needed link between singular and general causal statements.

We have noticed that the notion of kind, or similarity, is crucially relevant to the notion of disposition, to the subjunctive conditional, and to singular causal statements. From a scientific point of view these are a pretty disreputable lot. The notion of kind, or similarity, is equally disreputable. Yet some such notion, some similarity sense, was seen to be crucial to all learning, and central in particular to the processes of inductive generalization and prediction which are the very life of science. It appears that science is rotten to the core.

Yet there may be claimed for this rot a certain undeniable fecundity. Science reveals hidden mysteries, predicts successfully, and works technological wonders. If this is the way of rot, then rot is rather to be prized and praised than patronized.

Rot, actually, is not the best model here. A better model is human progress. A sense of comparative similarity, I remarked earlier, is one of man's animal endowments. Insofar as it fits in with regularities of nature, so as to afford us reasonable success in our primitive inductions and expectations, it is presumably an evolutionary product of natural selection. Secondly, as remarked, one's sense of similarity or one's system of kinds develops and changes and even turns multiple as one matures, making perhaps for increasingly dependable prediction. And at length standards of similarity set in which are geared to theoretical science. This development is a development away from the immediate, subjective, animal sense of similarity to the remoter objectivity of a similarity determined by scientific hypotheses and posits and constructs. Things are similar in the later or theoretical sense to the degree that they are interchangeable parts of the cosmic machine revealed by science.

This progress of similarity standards, in the course of each individual's maturing years, is a sort of recapitulation in the individual of the race's progress from muddy savagery. But the similarity notion even in its theoretical phase is itself a muddy notion still. We have offered no definition of it in satisfactory scientific terms. We of course have a behavioral definition of what counts, for a given individual, as similar to what, or as more similar to what than to what; we have this for similarity old and new, human and animal. But it is no definition of what it means really for *a* to be more similar to *b* than to *c*; really, and quite apart from this or that psychological subject.

Did I already suggest a definition to this purpose, metaphorically, when I said that things are similar to the extent that they are interchangeable parts of the cosmic machine? More literally, could things be said to be similar in proportion to how much of scientific theory would remain true on interchanging those things as objects of reference in the theory? This only hints a direction; consider for instance the dimness of 'how much theory'. Anyway the direction itself is not a good one; for it would make similarity depend in the wrong way on theory. A man's judgments of similarity do and should depend on his theory, on his beliefs; but similarity itself, what the man's judgments purport to be judgments of, purports to be an objective relation in the world. It belongs in the subject matter not of our theory of theorizing about the world, but of our theory of the world itself. Such would be the acceptable and reputable sort of similarity concept, if it could be defined.

It does get defined in bits: bits suited to special branches of science. It is in this way, on many limited fronts, that man continues his rise from savagery, sloughing off the muddy old notion of kind or similarity piecemeal, a vestige here and a vestige there. Chemistry, the home science of water-solubility itself, is one branch that has reached this stage. Comparative similarity of the sort that matters for chemistry can be stated outright in chemical terms, that is, in terms of chemical composition. Molecules will be said to *match* if they contain atoms of the same elements in the same topological combinations. Then, in principle, we might get at the comparative similarity of objects *a* and *b* by considering how many pairs of matching molecules there are, one molecule from *a* and one from *b* each time, and how many unmatching pairs. The ratio gives even a theoretical measure of relative similarity, and thus abundantly explains what it is for *a* to be more similar to *b* than to *c*. Or we might prefer to complicate our definition by allowing also for degrees in the matching of molecules; molecules having almost equally many atoms, or having atoms whose atomic numbers or atomic weights are almost equal, could be reckoned as matching better than others. At any rate a lusty chemical similarity concept is assured.

From it, moreover, an equally acceptable concept of kinds is derivable, by the paradigm-and-foil definition noted earlier in this paper. For it is a question now only of distilling purely chemical kinds from purely chemical similarity; no admixture of other respects of similarity interferes. We thus exonerate water-solubility, which, the last time around, we had reduced no farther than to an unexplained notion of kind. Therewith also the associated subjunctive conditional, 'If this were in water, it would dissolve', gets its bill of health.

The same scientific advances that have thus provided a solid underpinning for the definition of solubility in terms of kinds, have also, ironically enough, made that line of definition pointless by providing a full understanding of the mechanism of solution. One can redefine water-solubility by simply describing the structural conditions of that mechanism. This embarrassment of riches is, I suspect, a characteristic outcome. That is, once we can legitimize a disposition term by defining the relevant similarity standard, we are apt to know the mechanism of the disposition, and so by-pass the similarity. Not but that the similarity standard is worth clarifying too, for its own sake or for other purposes.

Philosophical or broadly scientific motives can impel us to seek still a basic and absolute concept of similarity, along with such fragmentary similarity concepts as suit special branches of science. This drive for a cosmic similarity concept is perhaps identifiable with the age-old drive to reduce things to their elements. It epitomizes the scientific spirit, though dating back to the Pre-Socratics: to Empedocles with his theory of four elements, and above all to Democritus with his atoms. The modern physics of elementary particles, or of hills in space-time, is a more notable effort in this direction.

This idea of rationalizing a single notion of relative similarity, throughout its cosmic sweep, has its metaphysical attractions. But there would still remain need also to rationalize the similarity notion more locally and superficially, so as to capture only such similarity as is relevant to some special science. Our chemistry example is already a case of this, since it stops short of full analysis into neutrons, electrons, and the other elementary particles.

A more striking example of superficiality, in this good sense, is afforded by taxonomy, say in zoology. Since learning about the evolution of species, we are in a position to define comparative similarity suitably for this science by consideration of family trees. For a theoretical measure of the degree of similarity of two individual animals we can devise some suitable function that depends on proximity and frequency of their common ancestors. Or a more significant concept of degree of similarity might be devised in terms of genes. When kind is construed in terms of any such similarity concept, fishes in the corrected, whale-free sense of the word qualify as a kind while fishes in the more inclusive sense do not.

Different similarity measures, or relative similarity notions, best suit different branches of science; for there are wasteful complications in providing for finer gradations of relative similarity than matter for the phenomena with which the particular science is concerned. Perhaps the branches of science could be revealingly classified by looking to the relative similarity notion that is appropriate to each. Such a plan is reminiscent of Felix Klein's so-called *Erlangerprogramm* in geometry, which involved characterizing the various branches of geometry by what transformations were immaterial to each. But a branch of science would qualify for recognition and classification at all, under such a plan, only when it had matured to the point of clearing up its similarity standards. Such branches of science would qualify further as unified, or integrated into our inclusive systematization of nature, only insofar as their several similarity concepts were *compatible*; capable of meshing, that is, and differing only in the fineness of their discriminations.

Disposition terms and subjunctive conditionals in these areas, where suitable senses of similarity and kind are forthcoming, suddenly turn respectable; respectable and, in principle, superfluous. In other domains they remain disreputable and practically indispensable. They may be seen perhaps as unredeemed notes; the theory that would clear up the unanalyzed underlying similarity notion in such cases is still to come. An example is the disposition called intelligence – the ability, vaguely speaking, to learn quickly and to solve problems. Sometime, whether in terms of proteins, colloids, nerve nets, or overt behavior, the relevant branch of science may reach the stage where a similarity notion can be constructed capable of making even the notion of intelligence respectable – and superfluous.

In general we can take it as a very special mark of the maturity of a branch of science that it no longer needs an irreducible notion of similarity and kind. It is that final stage where the animal vestige is wholly absorbed into the theory. In this career of the similarity notion, starting in its innate phase, developing over the years in the light of accumulated experience, passing then from the intuitive phase into theoretical similarity, and finally disappearing altogether, we have a paradigm of the evolution of unreason into science.[15]

Harvard University

REFERENCES

[1] C. G. Hempel, *Aspects of Scientific Explanation and Other Essays*, Free Press, New York, 1965, p. 15.
[2] Nelson Goodman, *Fact, Fiction, and Forecast*, University of London Press, London, 1955, p. 74.
[3] *Ibid.*, pp. 82 ff.
[4] *Ibid.*, pp. 95 ff.
[5] I mean this only as a sufficient condition of lawlikeness. See Donald Davidson, 'Emeroses by Other Names', *The Journal of Philosophy* **63** (1966), 778–780.
[6] This relevance of kind is noted by Goodman, first edition, pp. 119 ff.; second edition, pp. 121 ff.
[7] Nelson Goodman, *The Structure of Appearance*, 2nd ed., Bobbs-Merrill, New York, 1966, pp. 163 ff.
[8] Rudolf Carnap, *The Logical Structure of the World*, University of California Press, Berkeley, Calif., 1967, pp. 141–147. (German edition 1928.)
[9] See my *Word and Object*, M.I.T. Press, Cambridge, Mass., 1960, pp. 83 ff. for further discussion and references.
[10] See *ibid.*, pp. 90–95.
[11] This was noted by S. Watanabe on the second page of his paper 'Une explication mathématique du classement d'objets', in *Information and Prediction in Science* (ed. by S. Dockx and P. Bernays), Academic Press, New York, 1965.
[12] J. J. C. Smart, *Philosophy and Scientific Realism*, Routledge and Kegan-Paul, London, 1963, pp. 68–72.
[13] *Fact, Fiction, and Forecast, op. cit.*, pp. 95 ff.
[14] R. Carnap, 'Testability and Meaning', *Philosophy of Science* **3** (1936), 419–471; **4** (1937), 1–40.
[15] I am indebted to Burton Dreben and Nelson Goodman for helpful criticisms of earlier drafts.

JAAKKO HINTIKKA

INDUCTIVE INDEPENDENCE AND THE PARADOXES OF CONFIRMATION

One of the most fruitful kinds of contributions to philosophical discussion is the discovery of a new conceptual problem. The more intricate such a problem is, the more further discussion it is likely to prompt. One of the best examples of problems that have provoked an extensive literature is the paradox of confirmation ('raven paradox') to which Hempel called the attention of the philosophical community in his 'Studies in the Logic of Confirmation[1]. Rarely, if ever, has a single article given rise to as much philosophical discussion. (A good runner-up is the literature on Goodman's queer predicates.[2]) It seems to me that a volume honoring the work of Carl G. Hempel would be incomplete without a paper dealing with this indirect contribution of his to philosophical discussion.

I shall assume that the reader has some idea as to what the complex of problems that goes under the name 'paradox of confirmation' is supposed to be. Very briefly, the source of the problem is that on many reasonable assumptions concerning confirmation (including those that are incorporated in the usual systems of inductive logic) the different cells of one's classificatory system[3] are completely symmetric.[4] Thus, if we have a 2×2 classification (say the cross-classification of the members of our domain into ravens and non-ravens as well as into black and non-black objects) all the different kinds of observed individuals support a generalization equally, provided merely that they are compatible with it. (In some inductive logics, this holds only if the observed individuals eliminate precisely as many competing simpler generalizations.[5]) Thus, if we have observed all kinds of individuals in our 2×2 classification except non-black ravens, the generalization 'all ravens are black' will be supported as strongly by an observed white shoe or by the proverbial red herring (i.e., by any non-black non-raven) as by an observed black raven, which seems to be intuitively unacceptable.

In the literature, many observations are made which illuminate the causes of this embarrassing situation. It seems to me that there neverthe-

N. Rescher et al. (eds.), Essays in Honor of Carl G. Hempel.

less is not, and cannot be, any single observation that constitutes the 'solution' to the paradox in the sense of eliminating the need of further questions. Instead, several largely independent insights are needed to appreciate fully the situation. The most basic claim was in effect already made by Hempel in his original paper.[6] He suggested that the appearance of paradox is misleading, and altogether due to the particular examples used to illustrate the processes of confirmation. These examples are misleading in that we have a great deal of background information concerning the concepts involved which is not taken into account when the allegedly paradoxical result is derived which states that the different confirmatory instances are on a par with each other. The confirmation theories in which the alleged paradox arises are supposed to be applicable only when all our background information is built into explicitly formulated evidence-statements, which is not true about such examples as, e.g., the familiar raven illustration.

I tend to agree with this point, although it seems to me unlikely that all relevant background information can always be expressed in some given language fixed in advance. However, even if we grant Hempel his point, this does not settle the problems which arise in this connection. For then we immediately face the further questions: What precisely is the relevant background information? How is it to be formulated? From the above diagnosis it is at once seen that certain types of answers to these questions are likely to fail. For instance, even if any number of further predicates (over and above 'raven' and 'black') are taken into consideration, this does not remove the 'paradox', unless other changes are made as well. For whatever assumptions are made concerning these new predicates, most of them are not likely to eliminate the symmetry between the four cells of our 2×2 classification, which is the source of trouble. I suspect that this fact is connected with the difficulty of giving a fully satisfactory account of the 'paradox' which the literature on the subject exemplifies.

Perhaps the most plausible answer to the question as to what the additional information is that breaks the symmetry is to say that we have some background knowledge concerning the relative numbers (relative frequencies) of different kinds of individuals (vis-à-vis our 2×2 classification) which exist in the universe.[7] If it is assumed that the number of non-black non-ravens is much greater than that of ravens, an individual of which it is known that it is a raven is much likelier to be an exception from the generalization 'All ravens are black' than an individual of which it is known that it is a non-raven. This fact is construed as showing why the observation of a black raven (observing of a raven that it is black) contributes more to the confirmation of the generalization than the observation of a non-black non-raven (observing of a non-black object that it is not a raven).

I have no doubt that this is one type of background information which can create such asymmetries between the different cells of one's classification scheme as suffice to explain our ascription of greater confirmatory power to observations of black ravens as compared with observations of white shoes or red herrings.[8] However, I doubt very much whether this kind of background information is operative in all cases in which individuals belonging to the different cells of one's classificatory system exhibit marked differences in their power to confirm a generalization. My doubts are largely due to the following observations:[9]

(1) There seem to be definite counterexamples to the claim that background knowledge of relative frequencies is always the source of the relevant asymmetries. For instance, a generalization like 'no material bodies are infinitely divisible' seems to be completely unaffected by questions concerning immaterial entities, independently of what one thinks of the relative frequencies of material and immaterial entities in one's universe of discourse. In fact, a materialist or near-materialist is quite as apt to disregard all evidence concerning whatever immaterial, infinitely divisible entities there may exist as an idealist or near-idealist.

(2) If we are presupposing anything like the familiar Carnapian systems of inductive logic and if our domain is infinite, then the degree of confirmation of our generalization is identically zero and is not affected by any observations.[10] Thus, to make sense of the claim that the observation of a black raven contributes more to the confirmation of the generalization 'All ravens are black' than the observation of a white shoe because of our knowledge of relative frequencies, one has to assume either a non-Carnapian probability measure or else to operate with concepts like Carnap's notion of an instance confirmation. The former line of thought, insofar as its results have reached the print, does not seem to have produced viable solutions to the paradox. Moreover, insofar as the present paper attempts to present us with such a solution, it does

not rely on any knowledge of relative frequencies. The latter concept is subject to grave doubts on other counts.[11]

(3) If the relevant background information concerns relative frequencies, its effect is a matter of degree, and depends on the domain of individuals that one is assuming. It seems obvious to me that effects of the former kind can sometimes be observed. But there also seem to be other cases in which the confirmatory effect of the 'wrong' kinds of positive instances is strictly zero, and not just *very* small. Furthermore, their uselessness seems to be largely independent of the choice of the domain. For instance, one may ask whether 'All ravens are black' is confirmed *at all*, even to an infinitesimal degree, by white shoes or red herrings. Is a hypothesis concerning the structure (say) of the hydrogen atom always confirmed by observations of atoms of other kinds? Surely in *some* cases one feels that indirect confirmation of this kind is *strictly* useless. If one decides to restrict one's universe of discourse to birds, is a generalization concerning a ubiquitous species better confirmed by observations concerning members of other species than a generalization concerning a rare species? I cannot detect an intuitive inclination to assent to these statements though a positive answer is implied by a relative-frequency explanation of the paradox.[12]

Thus there seems to be plenty of room for attempts to find other kinds of background information, different from knowledge of relative frequencies, that might in some cases explain why we prefer one kind of confirmatory instance to another. The present paper strives to locate another possible source of asymmetries of the relevant kind. I shall present this partial solution to the paradox in a somewhat roundabout way. First, I shall set up the rudiments of a formal system of confirmation theory (inductive logic) in which the paradox does not arise. The fact that this can be done should at the very least dispel the confusions of those philosophers who seem to think that the paradox of confirmation is somehow intrinsic in an explicit logical treatment of confirmation theory. The systems of inductive logic which we shall be dealing with here are variants of the systems presented in my paper, 'A Two-Dimensional Continuum of Inductive Methods'.[13] Familiarity with this paper will be presupposed. Only after the formal or semi-formal method of avoiding the paradox has been expounded and some of its consequences have been studied shall we comment on the intuitive meaning of the assumptions it incorporates.

The results which we obtain in dealing with the paradox of confirmation have other applications as well. We shall briefly consider their relevance to the questions as to what happens when new primitive predicates are introduced into our system, or when quantifiers are restricted into a part of our universe of discourse only. Furthermore, the applicability of my methods to the problems which are usually formulated in terms of Goodman's queer predicates is also pointed out.[14]

Heuristically, the outlines of the problem are clear enough. The problem arises, as was already indicated, from the fact that all the different Q-predicates that can be formed from a number of given primitive predicates $P_1, P_2, \ldots, P_k$ are on a par. We cannot hope to deal with the paradox without creating some sort of asymmetry among them. One obvious way of doing this is to assume that the order of the primitive predicates somehow matters. This leads us to consider what shape our inductive logic might take if the predicates $P_1, P_2, \ldots, P_k$ are taken in this order and if this order has to be taken into account in some way or other.

One obvious way of taking it into account is to think of the Q-predicates as arising through a number of successive dichotomies: First all individuals are divided into the classes P_1 and $\bar{P}_1$, then each of these classes is divided into the classes of those individuals which have P_2 and of those which do not have it, and so on. Let us fix on some definite Q-predicate Ct_0 and let us designate by $n_0^{(i)}$ the number of those observed individuals which fall into the same class as members of Ct_0 in the i:th dichotomy (out of the totality of n observed individuals). Then it lies close at hand to think of the representative (characteristic) functions[15] of the different dichotomies. In terms of Carnap's λ-system, we are therefore bound to have something like

$$(1) \qquad \frac{n_0^{(1)} + \lambda_1/2}{n + \lambda_1} \quad \frac{n_0^{(2)} + \lambda_2/2}{n_0^{(1)} + \lambda_2} \quad \cdots \quad \frac{n_0^{(k)} + \lambda_k/2}{n_0^{(k-1)} + \lambda_k}.$$

Here it is assumed that the different dichotomies can be characterized by different parameters λ. It is natural, however, to put $\lambda = \lambda_1 = 2\lambda_2 = \cdots = = 2^{k-1}\lambda_k$.[16] Then (1) reduces to the old Carnapian form

$$(1') \qquad \frac{n_0 + (\lambda/K)}{n + \lambda},$$

where $K = 2^k$ and $n_0 = n_0^{(k)}$ is the number of observed individuals having the Q-predicate Ct_0 in question.

This reduction of (1) to (1′) shows that not very much new has been accomplished so far. However, the characteristic features of the two-dimensional continuum of inductive methods will help us here. We may take (1′) to be merely the conditional representative function which holds only on the condition that C_K be true.[17] In order to specify a system similar to the two-dimensional continuum we have to specify the other representative functions as well, one for each different constituent C_w, to whose truth it will be relative. Here of course $K>w>0$.

A relative representative function is obtained for this purpose in the same way (1) was obtained:

$$(2)\qquad \frac{n_0^{(1)}+\lambda_1/d_1}{n+\lambda_1}\ \frac{n_0^{(2)}+\lambda_2/d_1d_2}{n_0^{(1)}+\lambda_2/d_1}\cdots\frac{n_0^{(k)}+\lambda_k/(d_1d_2\ldots d_k)}{n_0^{(k-1)}+\lambda_k/(d_1\ldots d_{k-1})}.$$

Here $d_1=1$ or 2 depending on whether or not one of the classes P_1, $\bar{P}_1$ is always empty when C_w is true. In general, $d_i=2$ or 1 depending on whether the i:th dichotomy we make use of in arriving at Ct_0 is a real dichotomy when C_w is true or whether one of the classes involved is implied by C_w to be empty.

If we make the natural assumption $\lambda=\lambda_1=\lambda_2=\cdots=\lambda_k$ which corresponds to our earlier assumption used in going from (1) to (1′) and which we shall usually make in the sequel, (2) reduces to the form

$$(2')\qquad \frac{n_k+\lambda/(d_1d_2\ldots d_k)}{n+\lambda}.$$

Clearly (1′) is the special case of (2′) with $d_1=d_2=\cdots=d_k=2$.[18] We can write (2′) also in the form

$$(2'')\qquad \frac{n_0+\lambda/2^{\delta_0}}{n+\lambda},$$

where $n_0=n_0^{(k)}$ is the number of observed individuals having the Q-predicate Ct_0 and δ_0 is the number of true dichotomies we encounter in arriving at Ct_0 when the predicates $P_1, P_2,\ldots, P_k$ are taken in this order. For another Q-predicate Ct_i we shall call the corresponding numbers n_i and δ_i. They will of course depend both on Ct_i and the particular constituent C_w relative to which we are setting up the characteristic function.

We can now see that the usual paradox about the black ravens does not arise if we can assume that the two relevant predicates R (=raven) and B (=black) are taken in this order. We may use a system which is obtained by putting $\alpha=0$ as an example, for the different values of α do not make any essential difference for my present purpose. (After all, α only makes a difference to the *a priori* probabilities.) We shall be working on this assumption until further notice. Let us assume that our evidence e can be summed up by the following 2×2 contingency table:

	R	$\sim R$
B	n_{rb}	$n_{\bar{r}b}$
$\sim B$	$n_{r\bar{b}}$	$n_{\bar{r}\bar{b}}$

We shall put $n_{rb}+n_{r\bar{b}}=n_r$, $n_{\bar{r}b}+n_{\bar{r}\bar{b}}=n_{\bar{r}}$, etc., $n_r+n_{\bar{r}}=n=n_b+n_{\bar{b}}$.

If $n_{r\bar{b}}=0$ but the other three numbers $\neq 0$, there are two constituents C_3 and C_4 compatible with the evidence. The former says that the cell $R\bar{B}$ is empty in the whole universe (and not just in our evidence), while the latter says that it is instantiated outside our evidence.

The *a posteriori* probability of C_3, i.e., $p(C_3 \mid e)$, will now be by Bayes' theorem

$$p(C_3 \mid e) = \frac{p(e \mid C_3)}{p(e \mid C_3) + p(e \mid C_4)}.$$

Assuming that our universe of discourse is infinite, we have in virtue of the ordering assumptions we have made

$$\begin{aligned} p(e \mid C_3) &= \frac{\lambda/2}{\lambda} \times \frac{1+\lambda/2}{1+\lambda} \times \frac{2+\lambda/2}{2+\lambda} \times \cdots \times \frac{n_{rb}-1+\lambda/2}{n_{rb}-1+\lambda} \\ &\quad \times \frac{\lambda/4}{n_{rb}+\lambda} \times \frac{1+\lambda/4}{n_{rb}+1+\lambda} \times \cdots \times \frac{n_{\bar{r}b}-1+\lambda/4}{n_b-1+\lambda} \\ &\quad \times \frac{\lambda/4}{n_b+\lambda} \times \frac{1+\lambda/4}{n_b+1+\lambda} \times \cdots \times \frac{n_{\bar{r}\bar{b}}-1+\lambda/4}{n-1+\lambda} \\ &= \frac{\Gamma(\lambda)}{\Gamma(\lambda/2)\,[\Gamma(\lambda/4)]^2} \\ &\quad \times \frac{\Gamma(n_{rb}+\lambda/2)\,\Gamma(n_{\bar{r}b}+\lambda/4)\,\Gamma(n_{\bar{r}\bar{b}}+\lambda/4)}{\Gamma(n+\lambda)}, \end{aligned}$$

where Γ is the familiar gamma function of higher analysis.[19] By the same token we have

$$p(e \mid C_4) = \frac{\Gamma(\lambda)}{[\Gamma(\lambda/4)]^3} \times \frac{\Gamma(n_{rb} + \lambda/4)\,\Gamma(n_{\bar{r}b} + \lambda/4)\,\Gamma(n_{\bar{r}\bar{b}} + \lambda/4)}{\Gamma(n + \lambda)}$$

Hence we have on the basis of Bayes' theorem

$$p(C_3 \mid e) = \frac{1}{1 + \dfrac{\Gamma(\lambda/2)}{\Gamma(\lambda/4)} \times \dfrac{\Gamma(n_{\bar{r}b} + \lambda/4)}{\Gamma(n_{rb} + \lambda/2)}}.$$

As one can see, this depends only on n_{rb}, that is to say, only on the number of observed black ravens. This means that the paradox is avoided (on the assumptions we have made): our raven generalization is confirmed by ravens only, while white shoes become red herrings for the purposes of confirmation.

If n_{rb} is a large number, we have the following asymptotic expression for $p(C_3 \mid e)$ which is easily obtained from the preceding result and from Stirling's formula:

$$p(C_3 \mid e) = \frac{1}{1 + (n_{rb})^{-\lambda/4}\,\Gamma(\lambda/2)/\Gamma(\lambda/4)}$$

which indicates roughly how the choice of λ affects the degrees of confirmation of C_3.

A more general treatment is obtained by assuming that what we have to do is to partition the set of our primitive predicates $\{P_1, P_2, \ldots, P_k\}$ and then order the cells of this partition. For instance, let us assume that in this way we obtain

$$\langle\{P_1, P_2, \ldots, P_{i_1}\}, \{P_{i_1+1}, \ldots, P_{i_2}\}, \ldots, \{P_{i_{j-1}}, \ldots, P_k\}\rangle.$$

Then the representative function will give in the case of Ct_0

$$\frac{n_0 + \lambda/(d_1 d_2 \ldots d_j)}{n + \lambda} = \frac{n_0 + \lambda/D_0}{n + \lambda},$$

where $D_0 = d_1 d_2 \ldots d_j$, and where d_1 is the number of those Q-predicates which can be formed from $P_1, P_2, \ldots, P_{i_1}$ and which are allowed by C_w to be instantiated. One and only one of them (say Ct_0^1) is compatible with Ct_0. Then d_2 is the number of those Q-predicates which can be formed from $P_1, P_2, \ldots, P_{i_1+1}, \ldots, P_{i_2}$, which are allowed by C_w to be exemplified, and which are compatible with Ct_0. By obvious generalization, d_m is the number of those Q-predicates which (i) can be formed from $P_1, P_2, \ldots, P_{i_m}$, which (ii) are compatible with $Ct_0^{i_m-1}$, which (iii) are compatible with $Ct_0^{i_m-1}$, and which (iv) are allowed by C_w to be instantiated. Only one of them is compatible with Ct_0; it is called $Ct_0^{i_m}$.

Again, from C_K we get the Carnapian representative function, but for C_w with $w \neq K$ we have results different from the earlier ones, except when our partition is the improper one in which $j=1$, $i_1=K$. This special case is the case of the earlier 'unordered' system of inductive logic.

Even more general a system is obtained by taking all the possible partitions and orderings and giving each of them a 'weight' or a probability p. Then the combined relative representative function will obviously be

$$\frac{n_0 + \lambda \sum (p/D_0)}{n + \lambda},$$

where the sum is taken over all the different partitions and orderings.

In the case of the raven-black example we have three members in this sum, corresponding to the partitions and orderings $\langle\{R\},\{B\}\rangle$, $\langle\{B\},\{R\}\rangle$, and $\langle\{RB\}\rangle$. Let the weights of these be p_1, p_2, and p_3. If we make the same assumptions as before, we can then see that the degree of confirmation $p(C_3 \mid e)$ is

$$1/(1 + \gamma),$$

where

$$\gamma = \frac{\Gamma(\lambda(p_1/2 + p_2/4 + p_3/3))\,\Gamma(\lambda(p_1/4 + p_2/4 + p_3/3))}{(\Gamma(\lambda/4))^3}$$
$$\times \frac{\Gamma(\lambda(p_1/4 + p_2/2 + p_3/3))\;\Gamma(n_{rb} + \lambda/4)}{\Gamma(n_{rb} + \lambda(p_1/2 + p_2/4 + p_3/3))\,\Gamma(n_{\bar{r}b} + \lambda(p_1/4 + p_2/4 + p_3/3))}$$
$$\times \frac{\Gamma(n_{\bar{r}b} + \lambda/4)\,\Gamma(n_{\bar{r}\bar{b}} + \lambda/4)}{\Gamma(n_{\bar{r}\bar{b}} + \lambda(p_1/4 + p_2/2 + p_3/3))}$$

for it is easily seen that $p(e \mid C_4)$ has its earlier expression while $p(e \mid C_3)$

is now

$$\frac{\lambda(p_1/2 + p_2/4 + p_3/3)}{\lambda} \times \frac{1 + \lambda(p_1/2 + p_2/4 + p_3/3)}{1 + \lambda} \times \cdots$$

$$\times \frac{n_{rb} - 1 + \lambda(p_1/2 + p_2/4 + p_3/3)}{n_{rb} - 1 + \lambda} \times \frac{\lambda(p_1/4 + p_2/4 + p_3/3)}{n_{rb} + \lambda} \cdots$$

$$\times \frac{n_{\bar{r}b} - 1 + \lambda(p_1/4 + p_2/4 + p_3/3)}{n_b - 1 + \lambda} \times \frac{\lambda(p_1/4 + p_2/2 + p_3/3)}{n_b + \lambda} \cdots$$

$$\times \frac{n_{\bar{r}\bar{b}} - 1 + \lambda(p_1/4 + p_2/2 + p_3/3)}{n - 1 + \lambda}$$

$$= \frac{\Gamma(\lambda)\, \Gamma(n_{rb} + \lambda(p_1/2 + p_2/4 + p_3/3))}{\Gamma(\lambda(p_1/2 + p_2/4 + p_3/3))\, \Gamma(\lambda(p_1/4 + p_2/4 + p_3/3))}$$

$$\times \frac{\Gamma(n_{\bar{r}b} + \lambda(p_1/4 + p_2/4 + p_3/3))\, \Gamma(n_{\bar{r}\bar{b}} + \lambda(p_1/4 + p_2/2 + p_3/3))}{\Gamma(\lambda(p_1/4 + p_2/2 + p_3/3))\, \Gamma(n + \lambda)}.$$

By way of approximation, we thus obtain $p(C_3 \mid e) = 1/(1+\gamma)$ where γ is, apart from a factor independent of the evidence,

$$n_{rb}^{-a_1\lambda} \times n_{rb}^{-a_2\lambda} \times n_{rb}^{-a_3\lambda},$$

where

$$a_1 = p_1/2 + p_2/4 + p_3/3 - 1/4 = p_1/4 + p_3/12$$

$$a_2 = p_3/12$$

$$a_3 = p_2/4 + p_3/12.$$

If $p_1 = 0$, $p_3 = 1$, this reduces to the old non-ordered treatment. If $p_1 = 1$, $p_2 = p_3 = 0$, we have the results obtained earlier in this paper. If $p_1 = p_3 = 1/2$, $p_2 = 0$, we obtain $a_2 = a_3 = 1/6$, $a_1 = 2/3$. In this situation, the observation of a black raven contributes more to the confirmation of C_3 than the observation of a non-raven. This holds in general when $p_1 > p_2 > p_3$, and the more so the greater p_1 is as compared with the other two. Notice, incidentally, that we cannot obtain the situation characterized by $p_3 = 1$, $p_1 = p_2 = 0$ by suitably choosing p_1 and p_2 when $p_3 = 0$. In other words, the unordered case $p_3 = 1$ cannot be considered as any sort of weighted combination of the two differently ordered cases $p_1 = 1$ and $p_2 = 1$.

In our discussion so far it has been assumed, among other things, that

$0<\lambda<\infty$. If $\lambda\to\infty$ we obtain similar results. It is easily seen, in fact, that we now have $p(C_3 \mid e)=1/(1+\gamma)$ where

$$\gamma = b_1^{n_{rb}} \times b_2^{n_{\bar{r}b}} \times b_3^{n_{\overline{rb}}}$$

and where

$$b_1 = \frac{p_1/4 + p_2/4 + p_3/4}{p_1/2 + p_2/4 + p_3} = \frac{1}{1 + p_1/4 + p_3/12} = \frac{1}{1 + a_1}$$

$$b_2 = \frac{1}{1 + p_3/12} = \frac{1}{1 + a_2}$$

$$b_3 = \frac{1}{1 + p_2/4 + p_3/12} = \frac{1}{1 + a_3}.$$

When $p_1=1$, $p_2=p_3=0$, we obtain

$$p(C_3 \mid e) = \frac{1}{1 + (4/5)^{n_{rb}}}$$

which depends on the number n_{rb} of observed black ravens only.

We have assumed all through our discussion above simply that our universe is infinite, but that nothing further is known of its composition. This means, *inter alia*, that we have not discussed the possible effects of one's knowledge of the relative frequencies of different kinds of individuals in our universe. As was indicated above, taking them into account gives us another promising line of attack on the paradoxes and in fact serves as the basis of a well-known attempted solution of the paradox, proposed by Hosiasson-Lindenbaum, Pears, Mackie *et al.* However, for reasons already given, it seems to me that this second line of attack is not the only possible one, and that it does not give us a completely satisfactory treatment of the paradox. From the point of view of our present treatment, it is needed only if we cannot assume that p_1 is large in comparison with p_2 and p_3. If we can make this assumption in the raven case, our dismissal of the evidence of the non-ravens need not be based on any unconscious inference from the relative frequencies of ravens and black things in our universe.

What was just said illustrates my earlier point that there may be essentially different kinds of background knowledge that we have to evoke in different cases in order to account for the paradox of confirmation. In some cases, we have to do with the effects of our implicit knowledge of certain frequencies on our inductive thinking. This is a relatively trivial problem, which can be handled as soon as we can handle confirmation theory satisfactorily in other respects. On the other hand, it seems to me that deeper assumptions are often involved in the cases considered as paradoxical. These assumptions do not amount to prior knowledge of frequencies, but rather to *a priori* assumptions of the order and asymmetry among our primitive predicates. What these latter assumptions really amount to remains to be discussed.

An interesting question which we can ask here and which throws some light on the nature of the assumptions just mentioned is the following: What happens when a new primitive predicate, say R, is introduced in addition to the old ones $P_1, P_2, \ldots, P_k$? We shall assume that in the order we have been considering before R is made the *last* predicate, while no order obtains between the P's.

First consider $p(C_w)$ where C_w is one of the old constituents. We have after the adjunction $p(C_w)=\sum_i p(C_w^{(i)})$ where $C_w^{(i)}$ are all the new constituents compatible with C_w. If $\alpha=0$, each of them has the probability *a priori* equal to $1/2^{2K+1}$. Since there are obviously 3^{2K-w} of them, we have after the adjunction

$$p(C_w) = 3^{K-w}/2^{2K} = (3/4)^K \frac{1}{3^w}.$$

Earlier, we had $p(C_w)=1/2^K$. The ratio of the new value to the old one is therefore

$$(3/2)^K \times (1/3^w).$$

As for $p(e \mid C_w)$, we have now $\sum p(e \mid C_w^{(i)})\, p(C_w^{(i)} \mid C_w)$ where $C_w^{(i)}$ are all the different new constituents compatible with C_w. Here we have, because of the assumption that R will be the last predicate in an ordering, $p(e \mid C_w^{(i)})=$the old value of $p(e \mid C_w)=v_0$. Hence the new value of $p(e \mid C_w)$ is $\sum_i v_0 \times p(C_w^{(i)} \mid C_w)=v_0=$its old value.

Hence the only change which our inductive logic has undergone as far as earlier probabilities are concerned is a change in the *a priori* probabilities of constituents.

If α is not zero, our result concerning $p(e \mid C_w)$ still stands. What we now have concerning the *a priori* probabilities of constituents is the following:

The old value of $p(C_w)$ is proportional to

$$\frac{\lambda w/K}{\lambda}\,\frac{1+\lambda w/K}{1+\lambda}\cdots\frac{\alpha-1+\lambda w/K}{\alpha-1+\lambda}=\frac{\Gamma(\lambda)\times\Gamma(\alpha+\lambda w/K)}{\Gamma(\lambda w/K)\times\Gamma(\alpha+\lambda)}.$$

Asymptotically this is

$$\frac{\Gamma(\lambda)}{\Gamma(\lambda w/K)}\times\alpha^{\lambda((w/K)-1)}.$$

The new value is a sum of terms of the form

$$\frac{\Gamma(\lambda)}{\Gamma(\lambda w'/2K)}\times\alpha^{\lambda((w'/2K)-1)},$$

where $w \leqslant w' \leqslant 2K$. When α is very large, incomparably the largest of these terms is

$$\frac{\Gamma(\lambda)}{\Gamma(2w/2K)}\times\alpha^{\lambda((2w/2K)-1)}$$

which is therefore approximately the new value of $p(C_w)$. However, it is seen to be precisely the old value, which the new value therefore approximates when the index of caution α is large.

Thus the effect of the addition of a new predicate on earlier results has been seen to be very slight if this new predicate is made later than all the old ones in some ordering provided that $\alpha \neq 0$. Asymptotically, this effect vanishes when one's index of caution α grows.

A converse effect may be studied when the scope of all our quantifiers is restricted to the field of a predicate, say P_1, which is the first one in our ordering. We want to consider what happens to the generalization of the form

$$\begin{aligned}(3)\qquad &[(Ex)(P_1(x)\;\&\;G_1(x))\;\&\;(Ex)(P_1(x)\;\&\;G_2(x))\;\&\cdots\\ &\quad\&\;(Ex)(P_1(x)\;\&\;G_w(x))]\\ &\quad\&\;(x)\,[(P_1(x)\supset[G_1(x)\vee G_2(x)\vee\cdots\vee G_w(x)]],\end{aligned}$$

where $G_1(x), G_2(x), \ldots, G_w(x)$ are Q-predicates without $P_1(x)$.

What happens if our universe is restricted to individuals satisfying $P_1(x)$ while P_1 is omitted from our predicate base? We shall assume that $(Ex)\,P_1(x)$ is true.

Let us consider the raven case as an example. Given the same assumptions concerning our evidence as were made before,

(4) $(x)(R(x) \supset B(x))$

had earlier the probability

$$\frac{1}{1 + \dfrac{\Gamma(\lambda/2)}{\Gamma(\lambda/4)} \times \dfrac{\Gamma(n_{rb} + \lambda/4)}{\Gamma(n_{rb} + \lambda/2)}}.$$

It is easily checked that we obtain, if we restrict our universe to ravens only and consider just the one predicate B, the following probability for $(x)\,B(x)$:

$$\frac{1}{1 + \dfrac{\Gamma(\lambda)}{\Gamma(\lambda/2)} \times \dfrac{\Gamma(n_{rb} + \lambda/2)}{\Gamma(n_{rb} + \lambda)}},$$

where the only change as compared with the earlier expression is that λ has been multiplied by two. This can be generalized.

Perhaps the most straightforward way of seeing that it can is to observe that the probability that a given unknown member of the restricted universe satisfies the restricted generalization is obtained precisely in the same way as the probability that it satisfies the old unrestricted generalization (3), given the fact that it satisfies $P_1(x)$. The only change is that now we have to use the value 2λ instead of the old λ. The same goes for any finite number of unknown individuals, and hence for the generalization itself.

This observation can be related to those suggestions for the solution of the paradox of confirmation which turn on the idea that the generalization 'All ravens are black' somehow speaks of ravens only. For what they show is that on the assumptions which I have made – viz., that the predicates are ordered $\langle\{R\}, \{B\}\rangle$ – the scope of the generalization 'All ravens are black' can be restricted to ravens without making any difference to the inductive situation, except for the change of λ to 2λ.

Conversely, the requirement that it must be possible to restrict the scope of a generalization like (4) to those individuals which satisfy $P_1(x)$ without changing the inductive situation can be used to motivate an ordering of our basic predicate in such a way that $P_1(x)$ precedes all the others. I shall not discuss the details here; suffice it to say that if certain further assumptions are made, the ordering assumption can be derived from the requirement that scopes of generalizations can be restricted in the indicated way without upsetting our inductive logic.

In a sense, our treatment of the paradox of confirmation may thus be viewed as a formalization of that proposed solution of the paradox according to which the gist of the situation lies in the fact that the generalization 'All ravens are black' speaks only of ravens and does not say anything of non-ravens.[20] There is an important difference here, however, between what I am saying and what 'solutions' of this kind usually contain. Often, this restriction is somehow thought of as a matter of logic – part of what we choose to let the generalization mean. According to the view on which the present discussion is based, to assume an ordering among the basic predicate is to make an empirical assumption concerning the factual situation. If we are justified in assuming that the scope of the generalization 'All ravens are black' can be restricted to ravens, then this means that we have some outside information which we can rely on concerning the factual situation. The paradox arises from the fact that this information, which colors our spontaneous view of the situation, is not incorporated in the usual treatments of the inductive situation. The paradox is resolved by pointing out how this extra information, which cannot be formulated as an explicit premiss in the kind of (monadic first-order) language we are considering, can be taken into account.

Thus the proposed treatment of the paradox of confirmation represents a step away from a purely logical treatment of induction and towards some suitable form of the Bayesian position. The inductive policy one adopts in cases like the raven generalization does not depend only on the language one employs; it also depends on one's assumptions (one's background information) concerning the possible inductive priority of some of the basic predicates over others, as reflected by the kind of ordering assumptions we have discussed.

In this respect our treatment of the paradox of confirmation is, I am tempted to say, more Bayesian than the so-called 'Bayesian solution' of the same paradox. This term is sometimes applied to the attempted solution which turns on one's prior knowledge (or beliefs) concerning the relative frequencies of different kinds of individuals in one's universe and which we have already discussed.[21] One reason why it is called

Bayesian is that the prior assumptions this solution envisages are more varied and much more dependent on one's evaluation of the inductive situation than the language-determined prior probability distributions of the Carnapians. In both these respects, our treatment goes much further toward Bayesianism than the 'Bayesian solution'. What we have done is to show how information different from information concerning relative frequencies can be taken into account. This extra information cannot be expressed by any statement in the kind of language we are dealing with, but as the case always is with all explicit inductive logics, it will have to go together with some definite probability distribution (prior distribution) on state-descriptions.

What precisely is this extra information? An exhaustive answer would break the confines of this essay. A partial answer can nevertheless be gathered from what has been said, especially from the results concerning what happens when new predicates are introduced or when one is omitted (while one's universe is restricted). It was seen that if suitable ordering assumptions are made, these changes have relatively insignificant effects. This means, generally speaking, that our ordering assumptions can be viewed as assumptions concerning the inductive independence – in one of the possible senses of this term – of one's predicates vis-à-vis each other. When a predicate or a family of predicates precedes another, generalizations in terms of the former are largely unaffected by the introduction or omission of predicates which belong to the latter or otherwise follow it in our ordering. Conversely, restricting one's universe to a *Q*-predicate of an earlier family often has a relatively slight effect on generalizations which are formulated in terms of a later family of predicates and which only speak of individuals satisfying that *Q*-predicate. These are, in their different ways, symptoms of some sort of inductive independence, and our ordering assumption can hence be viewed as assumptions concerning the inductive independence of different predicates of each other. The precise nature of this independence would require further investigation.

This perhaps also helps us to appreciate the ease at which we seem to be able to make (in actual inductive situations) the assumptions which according to my suggestion amount to the ordering assumptions we have discussed. As Keynes in effect pointed out,[22] one important class of judgments we seem to be capable of making (though not always of articulating) with considerable accuracy are precisely judgments of inductive independence. In some cases at least, the independence in question seems to be of the kind we have discussed. The system of weighting different partitions and different orderings gives us a flexible method of expressing many different kinds of independence assumptions.

It is also of interest to see that certain problems Goodman has raised can be dealt with along the same lines as the raven paradox.[23] Here we shall deal with an example which is slightly different from those Goodman himself has employed but which seems to catch the gist of his objections to Carnapian inductive logics.

Let us assume that our domain of individuals consists of emerald-stages; that is, space-time points occupied by emeralds. The two basic predicates we have are the following:

$$G(x) = x \text{ is green}$$

$$O(x) = x \text{ occurs before 2000 A.D.}$$

In terms of these we can define the 'unnatural' predicate grot(x) as follows:

$$(x)\,[\text{grot}(x) = ((G(x)\ \&\ O(x)) \vee (\sim G(x)\ \& \sim O(x)))]\,.$$

In other words, an emerald-stage has the predicate grot if it either is green and occurs before 2000 A.D. or else is not green and occurs during or after 2000 A.D.

Goodman's puzzle can now be expressed by asking: Why is it that our observations of emerald-stages support the generalization 'All emerald-stages are green', symbolically

$$g_1 = (x)\,G(x)$$

and not the generalization g_2 = 'All emerald-stages are grot'. It is assumed here of course that all the members of our domain that have been observed have the Q-predicate

$$Ct_1(x) = (G(x)\ \&\ O(x))\,.$$

Furthermore, it is appropriate to assume that there will be emeralds in existence during or after 2000 A.D.; otherwise the green-grot distinction does not make any difference extensionally. In other words, the following statement is assumed to be true:

$$s = (Ex) \sim O(x)\,.$$

It is obvious that (in this particular case at least) all the talk of queer predicates is merely a way of dramatizing a certain problem which can be perfectly well posed and discussed without recourse to them.

In order to explain the problem, let us define the three last Q-predicates in the obvious way:

$$Ct_2(x) = G(x) \,\&\sim O(x)$$
$$Ct_3(x) = \sim G(x) \,\&\, O(x)$$
$$Ct_4(x) = \sim G(x) \,\&\sim O(x).$$

Since s is known to be true and since $Ct_1(x)$ has been observed to be instantiated, there remain six constituents compatible with s and with the evidence e. They can be specified by specifying which Q-predicates they assert to be exemplified. The following list thus serves to define them:

Constituent:	*Q-predicates exemplified:*
C_1	Ct_1, Ct_2
C_2	Ct_1, Ct_4
C_3	Ct_1, Ct_2, Ct_4
C_4	Ct_1, Ct_3, Ct_2
C_5	Ct_1, Ct_3, Ct_4
C_6	Ct_1, Ct_3, Ct_2, Ct_4.

Since the asymptotic behavior of degrees of confirmation does not depend on the *a priori* probabilities of the constituents, these may be assumed to be equal. On this assumption the degree of confirmation of C_i ($i=1, 2, \ldots, 6$) will be by the theorem of Bayes (with obvious simplifications)

$$p(C_i \mid e \,\&\, s) = \frac{p(e \mid C_i)}{p(e \mid C_1) + p(e \mid C_2) + \cdots + p(e \mid C_6)}.$$

Hence all that we have to do here is to calculate the likelihoods:

$$p(e \mid C_1), \ldots, p(e \mid C_6).$$

Now if all the different Q-predicates are on a par, it follows from the symmetry between Ct_2 and Ct_4 that $p(C_1 \mid e \& s) = p(C_2 \mid e \& s)$. However, it is readily seen that on the assumptions we have made, the truth of C_1 is tantamount to that of g_1 and the truth of C_2 to that of g_2. Thus

g_1 and g_2 are also equally probable on our evidence e, taken together with s. This, it seems to me, is the gist of Goodman's paradox. What was just said shows that logically speaking it is, like the paradox of confirmation, an almost direct consequence of the symmetry between the different Q-predicates.

In order to see how our version of Goodman's paradox can be treated along the same lines as the raven paradox, let us assume that our predicates are ordered as follows: $\langle\{G\}, \{O\}\rangle$ and calculate the different likelihoods. We obtain

$$p(e \mid C_1) = \frac{0 + \lambda/4}{0 + \lambda/2} \cdots \frac{n - 1 + \lambda/4}{n - 1 + \lambda/2} = \frac{\Gamma(\lambda/2)}{\Gamma(\lambda/4)} \times \frac{\Gamma(n + \lambda/4)}{\Gamma(n + \lambda/2)}$$

$$p(e \mid C_2) = \frac{0 + \lambda/2}{0 + \lambda} \cdots \frac{n - 1 + \lambda/2}{n - 1 + \lambda} = \frac{\Gamma(\lambda)}{\Gamma(\lambda/2)} \times \frac{\Gamma(n + \lambda/2)}{\Gamma(n + \lambda)}$$

$$p(e \mid C_3) = \frac{\Gamma(\lambda) \times \Gamma(n + \lambda/4)}{\Gamma(\lambda/4) \times \Gamma(n + \lambda)}$$

$$p(e \mid C_4) = p(e \mid C_3)$$

$$p(e \mid C_5) = p(e \mid C_2)$$

$$p(e \mid C_6) = p(e \mid C_3).$$

Here n is the number of observed emerald-stages. Thus

$$p(C_1 \mid e \;\&\; s) = \frac{p(e \mid C_1)}{p(e \mid C_1) + \cdots + p(e \mid C_6)}$$

$$= \frac{1}{1 + 2\dfrac{p(e \mid C_2)}{p(e \mid C_1)} + 3\dfrac{p(e \mid C_3)}{p(e \mid C_1)}}$$

$$\frac{1}{1 + 2\dfrac{\Gamma(\lambda)\Gamma(\lambda/4)}{[\Gamma(\lambda/2)]^2}\dfrac{[\Gamma(n + \lambda/2)]^2}{\Gamma(n + \lambda)\,\Gamma(n + \lambda/4)} + 3\dfrac{\Gamma(\lambda) \times \Gamma(n + \lambda/2)}{\Gamma(\lambda/2) \times \Gamma(n + \lambda)}}.$$

When n is large, this is by Stirling's formula approximately

$$\frac{1}{1 + 2\dfrac{\Gamma(\lambda)\,\Gamma(\lambda/4)}{[\Gamma(\lambda/2)]^2}\,n^{-(\lambda/4)} + 3\,\dfrac{\Gamma(\lambda)}{\Gamma(\lambda/2)}\,n^{-(\lambda/2)}}$$

which $\to 1$ when $n \to \infty$.

This means that the example of Goodman's paradoxes which we have been dealing with can be solved if we are entitled to order the two predicates *G*, *O* in the way we did. For what we have just seen shows that on this ordering assumption the generalization 'All emerald-stages are green' will become highly confirmed when our evidence grows while the generalization 'All emerald-stages are grot' will receive arbitrarily small degrees of confirmation when *n* grows.

A few words can be said at this stage of the relation of our proposed treatment of Goodman's puzzles to the way in which Goodman himself proposes to treat them. (I assume that the outlines of his approach are known to the reader.) There might seem to be a certain amount of convergence between our views. Goodman's treatment turns on considering certain predicates in a sense inductively better (more firmly 'entrenched') than others. This is shown (roughly speaking) by their past success in inductive generalizations. In the example at hand, 'grot' is weakly entrenched whereas 'green' is much better entrenched. As far as the basic predicates are concerned, 'occurs before 2000 A.D.' is a weakly entrenched predicate, as might be expected from the fact that the loosely entrenched predicate 'grot' can be defined in terms of it and 'green'. These comparisons between different basic predicates might be compared with the ordering of predicates on which my treatment was based. Is there perhaps some deeper reason for the fact that my treatment was based on an ordering which matches the relative degrees of entrenchment of the two predicates?

There may be some actual connections here, but it seems to me that they are largely accidental. If I am right in thinking that Goodman's paradox is essentially similar in its logical structure to the paradox of confirmation, then we should be able to say the same things of the latter as of the former. Now my treatment of the 'raven paradox' turned on considering the two predicates 'raven' and 'black' in this order. Does this mean that the predicate 'raven' is more firmly entrenched than 'black'? I am not sure whether I understand fully Goodman's explanations of the notion of entrenchment, but insofar as I grasp them, this does not seem to be the case according to Goodman's explanations. For an important factor in the entrenchment of a predicate is (*apud* Goodman) its history of successes in past inductions ('projections'). Now surely the term 'black' figures in many more successful past inductions than the term 'raven'.[24] Hence it ought to be more firmly entrenched than 'raven' and precede it in the corresponding ordering, although we saw that the precise opposite ordering was required to handle the raven paradox. Our ordering thus cannot be identified in general with an ordering that depends on relative degrees of entrenchment in Goodman's sense. It seems to me preferable to try to look at the orderings in terms of assumptions of inductive independence.

University of Helsinki and Stanford University

REFERENCES

[1] Carl G. Hempel, 'Studies in the Logic of Confirmation', *Mind* **54** (1945), 1–26 and 97–121; reprinted in Carl G. Hempel, *Aspects of Scientific Explanation and Other Essays*, The Free Press, New York, 1965, pp. 3–46. Although the problem had been raised before Hempel, virtually the whole subsequent discussion seems to have derived its impetus from his article. I shall not try to survey this subsequent discussion of the paradox. Excellent partial surveys are provided by Israel Scheffler in *The Anatomy of Inquiry*, A. A. Knopf, New York, 1963, pp. 241–242, 258–295, and by Max Black in 'Notes on the Paradoxes of Confirmation', in *Aspects of Inductive Logic* (ed. by Jaakko Hintikka and Patrick Suppes), North-Holland Publ. Co., Amsterdam, 1966, pp. 175–197.

[2] See Nelson Goodman, *Fact, Fiction, and Forecast*, University of London Press, London, 1955.

[3] If we start from a number of primitive (monadic) predicates P_1, P_2, ..., P_k, these cells are defined by the complex predicates which Carnap calls *Q*-predicates and which specify which primitive predicates belong to an individual. They are thus of the form

$$(\pm)\, P_1(x) \,\&\, (\pm)\, P_2(x) \,\&\, \cdots \,\&\, (\pm)\, P_k(x)$$

where each $(\pm)$ can be replaced by a negation sign or it can also be omitted in all the different combinations. For *Q*-predicates and related matters, see Rudolf Carnap, *Logical Foundations of Probability*, University of Chicago Press, Chicago, 1950 (second ed., 1962); *The Continuum of Inductive Methods*, University of Chicago Press, Chicago, 1952.

[4] This holds *inter alia* of all the members Carnap's λ-continuum of inductive methods. (See the works listed in reference 3.)

[5] Cf. my paper, 'Induction by Elimination and Induction by Enumeration', in *The Problem of Inductive Logic* (ed. by I. Lakatos), North-Holland Publ. Co., Amsterdam, 1968.

[6] Hempel, *op. cit.* Cf. also Scheffler's and Black's discussions of Hempel's point, *op. cit.*, pp. 258–269 and pp. 187–188, respectively.

[7] This kind of solution of the 'paradox' has been put forward repeatedly; see e.g., J. Hosiasson-Lindenbaum, 'On Confirmation', *The Journal of Symbolic Logic* **5** (1940), 138–148; D. Pears, 'Hypotheticals', *Analysis* **10** (1950), 49–63; J. L. Mackie, 'The Paradoxes of Confirmation', *The British Journal for the Philosophy of Science* **13** (1963),

265–277; Patrick Suppes, 'A Bayesian Approach to the Paradoxes of Confirmation', in *Aspects of Inductive Logic, op. cit.*, pp. 198–207.

[8] One thought-experiment which illustrates the effects of relative frequencies might be the following: Suppose our 'population' consists of all patients suffering from a certain illness, and suppose that we are considering the generalization 'All patients who are given the drug x are cured'. If relatively few patients are administered the drug, we are naturally interested in what happens to them. But assume that the use of the drug becomes almost universal, and yet no counter-examples are produced. Would we not become increasingly more and more interested in those relatively few cases in which a cure is known not to have obtained, and inquire into whether *these* patients had been given the drug or not?

[9] My doubts are closely related to the objections which Max Black puts forward in *op. cit.*, pp. 195–197 against the same line of thought (which he calls the 'Bayesian solution' to the paradox).

[10] See Carnap's works listed in reference 3.

[11] See e.g., my paper 'Towards a Theory of Inductive Generalization', in *Logic, Methodology, and Philosophy of Science*, Proceedings of the 1964 International Congress (ed. by Yehoshua Bar-Hillel), North-Holland Publ. Co., Amsterdam, 1965, pp. 274–288.

[12] If this point is correct, then a satisfactory probabilistic treatment of such paradoxical situations as the raven case (supposing that it follows this pattern) must yield the result that the *a posteriori* probability of the generalization 'All ravens are black' depends *only* on the number of observed ravens. We shall outline one treatment of this kind.

[13] 'A Two-Dimensional Continuum of Inductive Methods', in *Aspects of Inductive Logic, op. cit.*, pp. 113–132.

[14] Goodman, *Fact, Fiction, and Forecast, op. cit.*

[15] In *A Continuum of Inductive Methods, op. cit.*, Carnap calls these functions characteristic functions, and in some of my earlier papers I have followed this practice. He now prefers to use the term 'representative function', however, in order to avoid clashes with the standard mathematical and statistical terminology, and I am glad to follow suit. (Carnap's book should be consulted for the details of his λ-system.)

[16] One way of seeing this naturalness is to recall the familiar 'interpretation' of Carnap's λ as the cardinality of a fictional sample of imaginary individuals evenly divided between the Q-predicates. Each dichotomy will naturally cut this 'sample' into half.

[17] If we follow the letter of 'A Two-Dimensional Continuum of Inductive Methods', *op. cit.*, we have to speak of 'truth is principle' here instead of truth *simpliciter*. However, the approach sketched there is easily modified in this respect, and in any case the difference does not affect the case of an infinite domain.

[18] Thus the simple suggestion we are following yields something essentially new only in connection with a treatment of induction in which generalizations receive non-zero *a priori* probabilities (even in an infinite universe). The fact that this does not happen to generalizations in Carnap's λ-system may have something to do with the fact that our treatment of the paradoxes has no precedent in the literature.

[19] In case you are not familiar with it, think of it as a generalization of the factorial $1 \times 2 \times 3 \times \cdots \times n = n!$ In fact, for a natural number n, $\Gamma(n) = (n-1)!$ Many of the familiar results on factorials (e.g., Stirling's formula) hold for Γ.

[20] Cf. e.g., G. H. von Wright, 'The Paradoxes of Confirmation'. in *Aspects of Inductive Logic, op. cit.*, pp. 208–218.

[21] Cf. e.g., P. Suppes, 'A Bayesian Approach...', *op. cit.*; M. Black, 'Notes...', *op. cit.*
[22] J. M. Keynes, *A Treatise on Probability*, Macmillan, London, 1921, pp. 255–256.
[23] *Fact, Fiction, and Forecast*, *op. cit.*
[24] We can obtain an even clearer example by taking (instead of the raven generalization) a generalization concerning a recently discovered species which has not yet figured in many inductions.

WESLEY C. SALMON

PARTIAL ENTAILMENT AS A BASIS FOR INDUCTIVE LOGIC

I. INTRODUCTION

Ever since the famous 'H_2O papers' of the early forties,[1] Carl G. Hempel has been closely associated with the concepts of confirmation and logical probability. His intriguing 'paradoxes of confirmation' have brought forth an avalanche of responses, and more keep coming.[2] Some of the conditions of adequacy he enunciated, especially the 'equivalence condition', are central to current controversies.[3] Moreover, he has, in the meantime, continued to make significant contributions to inductive logic,[4] as well as to such closely allied topics as the nature of theories and inductive (statistical) explanation.[5]

In the early days, Helmer-Hempel-Oppenheim maintained that the qualitative concept of *confirmation* as well as the quantitative concept of *degree of confirmation* could be characterized in a purely syntactical manner. Hempel, at least, has subsequently been convinced by Nelson Goodman's 'grue-bleen paradox' that a purely syntactical definition is impossible.[6] While he points out that Goodman's own solution via 'entrenchment' is obviously pragmatic, Hempel does not commit himself to the thesis that confirmation and degree of confirmation are inescapably pragmatic. Personally, I think it is possible to provide a straightforwardly semantical solution to Goodman's paradox by imposing certain conditions upon the interpretation of primitive descriptive predicates.[7] In this way we can avoid making confirmation a pragmatic relation. If such an approach to Goodman's paradox is viable, it leaves open the possibility that confirmation and degree of confirmation are logical relations, in the broad sense of 'logical' that encompasses both syntactical and semantical considerations. This is the sense of 'logical' that will be employed in the following discussion.

The purpose of this paper will be to examine the concept of logical probability with a view to considering its suitability as a fundamental *logical* relation upon which to ground inductive logic. A good deal of

N. Rescher et al. (eds.), Essays in Honor of Carl G. Hempel.

attention will be focused upon the relation of *partial entailment* which has sometimes been used to explain the concept of logical probability. The basic thesis I wish to examine is that the logical status of induction can be understood in terms of a strong analogy with deduction, according to which inductive logic rests upon a relation of partial entailment just as deductive logic rests upon a relation of full logical entailment. Taking *deductive entailment* with its correlative *logical independence* as a point of departure, I shall show how a notion of partial entailment arises naturally within the context of deductive logic, and I shall introduce a relation of *complete independence*, quite distinct from logical independence, which obtains when there is neither full (deductive) entailment nor partial entailment.[8] As a matter of fact, when it comes to providing formal definitions, it will be easier to define complete independence first. This relation can then be used to help clarify partial entailment. In the course of the discussion, the concept of (semantic) *content*, which at first blush might seem to hold promise of providing a simple basis for the definition of complete independence, will be considered. I shall claim that the concept of content has no particular utility along these lines, but our discussion of complete independence will lead to a useful concept of *independence of content*.

In the end I shall be arguing that full clarity on the foregoing concepts, and the relations among them, militates against taking logical probability (or any other *logical* relation) as the basic relation for inductive logic. The main difficulty is that the notion of partial entailment arising naturally out of an examination of deductive relations is radically incongruous with the relation of partial entailment needed to serve as the fundamental relation of inductive logic.[9] This is not to say that it is impossible to define an alternative concept of partial entailment for the purposes of inductive logic – clearly it is possible to do so – but the incongruity removes all plausibility from the analogy between deduction and induction as logical systems based upon the relations of full and partial entailment respectively. If there is serious question about the status of the logical relation upon which inductive logic is supposed to rest – and how can there fail to be? – it is not answered by this analogy. I shall be arguing, as it develops, that the incongruity between the two distinct relations of partial entailment – that arising naturally from deductive logic and that needed for inductive logic – rests upon the extremely fundamental and sometimes unnoticed distinction between degree of *confirmation* and degree of *relevance* (Carnap's 'concepts of firmness' and 'concepts of increase of firmness').[10]

This essay is, I must hasten to emphasize, largely an exercise in heuristics. The concept of partial entailment has not, I believe, been used directly in the explication of the relation of logical probability upon which inductive logic is to be erected, but rather in the course of attempts to clarify the explicandum. In this role it has borne a large part of the burden of lending plausibility to the whole enterprise of attempting to ground inductive logic on a logical relation. It is my conviction that a close examination of the salient concepts removes the plausibility of this undertaking, and lends much greater plausibility to the view that the only relations upon which inductive logic can be said to rest are factual. The outcome, I believe, is entirely Humean: the unobserved is logically independent of the observed, and this means that neither deductive nor inductive logical relations provide the bridge from the observed to the unobserved needed to ground inductive logic.

II. THE LANGUAGES L_N

Inasmuch as Carnap has provided precise foundations for both deductive and inductive logic, I shall utilize his terminology and his apparatus.[11] In fact, I shall confine my attention to his finite languages L_N. These have the structure of first order applied functional calculi with identity, and each has a finite number of primitive predicates and a finite number N of individual constants. Such languages are interpreted over finite domains of N individuals. The restriction to such simple languages is a severe curtailment of generality, but it will still be possible to elicit the fundamental issues.

In the languages L_N, an *atomic statement* is one that consists of a primitive predicate followed by an appropriate number of individual constants. Any truth-functional compound of atomic statements is a *molecular statement*. A *basic statement* is an atomic statement or the negation of an atomic statement. An atomic statement together with its negation is a *basic pair*. A *state description* is a conjunction which contains as conjuncts one and only one member of each basic pair. For the languages L_N the notion of a statement *holding in* a state description can be defined precisely.[12] The *range* $R(p)$ of a statement p in L_N is the set of all state descriptions of L_N in which p holds. Since such relations as entailment, consistency, and incompatibility are defined in terms of ranges a flagrant circularity would arise if they were used to *define* 'holding in', but it is an immediate consequence of the definitions that a statement, holds in any state description with which it is compatible, and it does not hold in any state description with which it is inconsistent. The range of a self-contradictory statement is null, but all consistent statements have nonempty ranges. The basic logical concepts, such as L-truth, L-implication, L-equivalence, L-exclusiveness, and L-disjunctiveness, are defined in terms of ranges; for example, p is L-true if $R(p)$ is the set of all state descriptions, p L-implies q if $R(p)$ is contained in $R(q)$, and p and q are L-exclusive if $R(p\&q)$ is null.[13] Any consistent statement can be expressed as (is L-equivalent to) the disjunction of all of the state descriptions in its range (where a state description standing alone constitutes a disjunction with only one disjunct). The standard concept of entailment, I take it, coincides with L-implication; at any rate, I shall construe 'entailment' in that way.

The languages L_N have essentially only a truth-functional structure, for every statement can be expressed as a truth function of atomic statements. Quantifiers can always be eliminated in favor of conjunctions or disjunctions.

III. ENTAILMENT AND INDEPENDENCE

Since the main purpose of this paper is to consider the possibility of grounding inductive logic in a logical relation between evidence and hypothesis, it seems reasonable to begin by considering the logical relations to be found in deductive logic, about which we have fairly clear conceptions. The central relation is that of entailment, in terms of which deductive validity can be understood. This relation provides the basis for correct deductive inference. If two statements do not have entailment relations between them, we say that they are *logically* – i.e., deductively – *independent* of one another.

If inductive logic is to be grounded in a logical relation, it is natural to look for a relation that is somehow analogous to entailment to serve as a basis for correct inductive inference. Inductive logicians have felt that such a relation exists, and it has been called 'partial entailment' or 'partial implication'.[14] This relation is weaker than full logical entailment, but it does not constitute *complete* logical independence.

As a matter of fact, examination of the logical relations among statements does reveal a relation that looks like a plausible candidate for the title of 'partial entailment'. In order to clarify the notion of partial entailment, we shall need to introduce a concept of complete independence. There seem to be two extremes, logical entailment and complete independence, and between them lies the relation of partial entailment. Complete independence is a narrower concept than logical independence; logical independence obtains when there are no full entailment relations, while complete independence demands that there be no full or partial entailments. Definitions of 'complete independence' and 'partial entailment' arise very naturally within the context of deductive logic, and it is easy to find paradigms of these relations.

The trouble that beclouds this apparently happy situation is that the relation of partial entailment that arises naturally out of the general consideration of deductive logical relations is radically different from the concept of partial entailment that inductive logicians have offered. In fact, they are basically incompatible with one another. For one thing, they differ on the very admissibility of the relation of complete independence. The relation that arises from deductive considerations demands that some statements stand in the relation of complete independence; there are clear examples of partial entailments and clear examples of complete independence. The relation inductive logicians have supplied, by contrast, excludes the very possibility of the relation of complete independence ever obtaining.

The relation of complete independence will play a central role in the subsequent discussion for several reasons. First, as we have just seen, it is a focal point of difference between the two concepts of partial entailment. Second, it will actually prove convenient to define 'complete independence' before 'partial entailment'. Third, and most importantly, in seeking a logical relation to underlie inductive logic, we are looking for a relation that embodies inductive dependence. By way of contrast, we shall need to be clear on what constitutes complete independence – that is to say, lack of any inductive or deductive dependency relation. As we shall see, the concept of logical independence can be characterized quite straightforwardly; the concept of complete independence is somewhat trickier. However, it is essential that we make a clear distinction between the two types of independence.

Let me now attempt to give some substance to these remarks by clarifying the main concepts with which we are concerned. The basic notion is logical entailment, which I take to be adequately explicated for present purposes by Carnap's concept of L-implication. Then we may say,

> p and q are *logically independent* $=_{\text{df}}$ none of the following relations holds: p entails q, p entails $\sim q$, $\sim p$ entails q, $\sim p$ entails $\sim q$.

In other words, two statements are logically independent unless a truth-value of one of them determines a truth-value for the other.

Suppose that p, q, r are three distinct atomic statements. In that case there will be no entailment relations between $p \& q$ and $q \& r$, but it is reasonable to say that there is some sort of logical dependency between them. Since they share a conjunct it seems appropriate to say that the relation is one of partial entailment. Furthermore, even in certain cases in which we do not have logical independence, it still seems natural to say that we have partial entailment, for we may have full entailment in one direction and partial entailment in the other. If p entails q but q does not entail p, we are inclined to say that q partially entails p, for q asserts part of what p asserts. One exception should perhaps be noted; if q is L-true it might be denied that q asserts anything at all, so we might be reluctant to say that it asserts part of what p does.

Since inductive logicians have often maintained that the truth of deductive consequences can lend inductive support to a hypothesis, the temptation is particularly strong to claim that q partially entails p whenever (but not necessarily only whenever) q is not L-true and p entails q. For instance, in L_3 let p be $(x)\, Fx$, which is equivalent to $Fa \& Fb \& Fc$, and let q be Fa. In this case also, q entails a conjunct of a statement L-equivalent to p.[15] In these cases we have found a deductive relation of dependency which falls short of full entailment, but which does not represent complete independence either. It is *one* plausible candidate for partial entailment.

Entailment relations can, as previously mentioned, be explicated in terms of ranges, i.e., p entails q if and only if $R(p)$ is included within

$R(q)$. If $R(p)$ overlaps $R(q)$ without being included in it, or if $R(q)$ is included within $R(p)$ but not conversely, it is sometimes said that p partially entails q. This is *another* concept of partial entailment that has seemed attractive to inductive logicians. Such relations among ranges are often presented diagrammatically by a device such as Euler circles, where the inclusion, exclusion, and overlap relations are shown pictorially.[16] At this point we must guard against an elementary confusion. When the ranges of p and q are mutually exclusive we do *not* have a relation of logical independence; rather, p entails $\sim q$ and q entails $\sim p$, so a full entailment holds in both directions. Thus, if we construe partial entailment as overlap of ranges, it follows immediately that between any two statements, going either way, there is either a full entailment or a partial entailment. The relation of complete independence, which is to consist in the absence of both full and partial entailments, never obtains.

These two incompatible concepts of partial entailment seem to arise because we have one picture of *logical relations* running from full entailment, through partial entailment, to complete independence, and another picture of *relations among ranges* extending from complete inclusion, through partial overlap, to complete exclusion. Each of these pictures

Logical relations	*Relations of ranges*
Full entailment	Inclusion
Partial entailment	Overlap
Complete independence	Exclusion

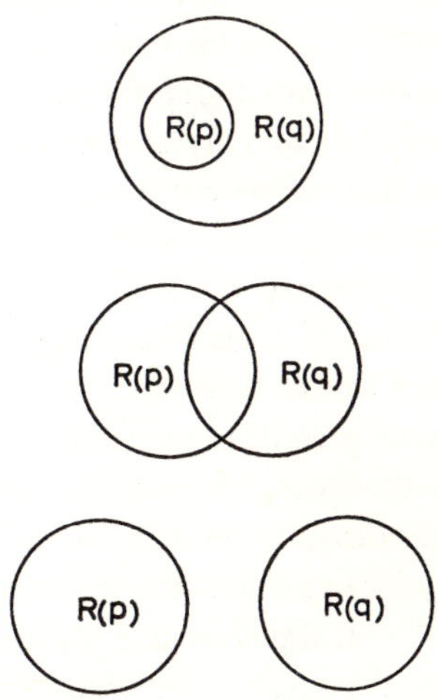

Fig. 1. Two incongruous scales.

is appealing, but they are mutually incongruous (see Figure 1). In particular, if we insist upon the possibility of complete independence, then we shall have to abandon the inductive logicians' approach that identifies partial entailment with overlap of ranges. Complete independence will have to be found in cases in which the ranges overlap, not in cases of mutually exclusive ranges. At this point it is not tempting to give up the concept of complete independence, for our paradigms point to three distinct relations: if p, q, and r are three distinct atomic statements, (i) $p \& q$ entails p, (ii) $p \& q$ partially entails $q \& r$, while (iii) p is completely independent of q. Hence, it is not clear that we should abandon the concept of partial entailment that allows room for complete independence.

In the sequel, I shall attempt to show how definitions of 'complete independence' and 'partial entailment' arise very naturally within the context of deductive logic. The trouble is that they are of no use in founding an inductive logic. Partial entailments of this sort fail to obtain just where we most need them in inductive logic. An inductive logic founded upon this concept of partial entailment would be transparently inadequate. I shall not argue, however, that this is the only possible definition of 'partial entailment'. As we have already seen, inductive logicians seem to have another concept in mind; indeed, this alternative becomes a vast set of alternatives when we consider the variety of ways in which overlap of ranges can be measured. Some of these alternatives appear to be far more suitable for inductive logic than is the concept that arises out of our consideration of deductive relations. However, if the concept of partial entailment is to provide a rationale for holding that inductive logic is based upon a logical relation, it must be a concept that arises naturally out of a consideration of logical relations. Otherwise, the burden is reversed. If a relation of partial entailment suitable for inductive logic does not emerge from an examination of logical relations, but rather, is extracted from a characteristically inductive concept such as degree of confirmation, then inductive logic must bear the burden of establishing the significance of the relation of partial entailment. Partial entailment can bear none of the burden of clarifying the foundations of inductive logic. But above all, we must avoid being lulled by the existence of a straightforward logical relation of partial entailment into supposing that we are thereby assured of the existence of a relation of partial entailment that can serve as a basis for inductive logic, for two distinct relations of partial entailment are involved, and they differ from one another in fundamental ways.

IV. CONTENT

As we have seen, the relation between statements with mutually exclusive ranges, far from being one of independence, is a relation of mutual logical entailment. It is initially tempting to suppose, however, that the concept of content will prove more useful than the notion of range in getting at the distinctions that concern us. For it certainly seems obvious, in the light of our simple paradigms of full and partial entailment, that the pairs of statements that sustain such relations share common content. Indeed, it is easy to suppose that these full or partial entailment relations obtain *by virtue of* shared content, so that lack of common content would characterize the relation of complete independence we have been seeking. I shall argue, nevertheless, that the concept of common content is actually of no use at all for this purpose, but in order to do so I shall, of course, need a reasonably precise account of the content of statements. Fortunately, one is available.

There are two notions of content which have strong preanalytic appeal.[17] First, it seems reasonable to think of the content of a statement p in terms of the statements p entails. In order to elucidate the content of a statement, one can exhibit its implications; indeed, we often characterize the very process of deductive inference as one which makes explicit the implicit content of the premisses. Thinking along these lines, it is plausible to identify the content of a statement with the set of all of the statements it entails. Second, it is natural to think of the content of a statement in terms of the possible states of affairs it excludes. For instance, we feel that tautologies have no content precisely because they are compatible with any possible state of affairs; they rule nothing out. Statements with factual content, in contrast, are incompatible with some states of affairs, and the stronger the factual statement, the greater the number of possible states of affairs with which it is incompatible. According to this second conception of content, the greater the content of a statement the more completely it narrows down the range of possible states of affairs which could be actual if that statement were true.

The foregoing notions of content, despite possible appearances to the contrary, are not in conflict with one another; they fit together harmoniously in terms of an account of *semantic content* given by Carnap and Bar-Hillel.[18] This explication is provided for the simple finite languages L_N with which we are working. In such languages, the state descriptions, which are *conjunctions* containing one member from each basic pair of statements, are the *strongest* consistent statements in the language. Any consistent statement is equivalent to the disjunction of the state descriptions in its range, and it is entailed by any such state description. Since every consistent statement has a nonempty range, every consistent statement is entailed by some state description. Analogously, it is possible to define the set of *weakest* factual statements in such a language. Such statements are simply *disjunctions* containing as disjuncts one statement from each basic pair. These are weakest statements in the sense that every non-L-true statement in the language entails at least one such. Any factual statement is equivalent to the conjunction of all of these weakest statements it entails. If we think of content in the first of our two ways, in terms of what a statement entails, then it is natural to characterize the content of statements in terms of these weakest statements just as the range of a statement is characterized in terms of the strongest statements (state descriptions). Indeed, since these disjunctions are weakest statements, we may think of them as having minimal content – as being content-atoms so to speak – or 'content elements' as Carnap and Bar-Hillel call them. According to this conception of content, then, the content of a factual statement of L_N is the set of content elements it entails.

Each content element is, of course, the negation of a state description. In L_3 with one predicate F we have:

	State descriptions		*Content elements*
1.	$Fa \,\&\, Fb \,\&\, Fc$	1.	$\sim Fa \vee \sim Fb \vee \sim Fc$
2.	$Fa \,\&\, Fb \,\&\, \sim Fc$	2.	$\sim Fa \vee \sim Fb \vee Fc$
3.	$Fa \,\&\, \sim Fb \,\&\, Fc$	3.	$\sim Fa \vee Fb \vee \sim Fc$
4.	$\sim Fa \,\&\, Fb \,\&\, Fc$	4.	$Fa \vee \sim Fb \vee \sim Fc$
5.	$Fa \,\&\, \sim Fb \,\&\, \sim Fc$	5.	$\sim Fa \vee Fb \vee Fc$
6.	$\sim Fa \,\&\, Fb \,\&\, \sim Fc$	6.	$Fa \vee \sim Fb \vee Fc$
7.	$\sim Fa \,\&\, \sim Fb \,\&\, Fc$	7.	$Fa \vee Fb \vee \sim Fc$
8.	$\sim Fa \,\&\, \sim Fb \,\&\, \sim Fc$	8.	$Fa \vee Fb \vee Fc$

Identification of the content of p with the content elements it entails is precisely equivalent to identifying that content with the set of state descriptions with which p is incompatible, that is, with the range of $\sim p$. Since the state descriptions are intended to be complete descriptions of possible states of the miniature universe describable by our language, the identification of the content of p with the statements p entails comes to exactly the same thing as identification of the content of p with the possible states of affairs incompatible with p.

Note, incidentally, that any state description is incompatible with all other state descriptions; one reason, therefore, for saying that state descriptions have maximal content is that each one specifies the state of the universe completely by excluding all other possible states. (The only statement stronger than a state description is an L-false statement; such a statement is incompatible with every state description and entails every content element. The only statement weaker than a content element is a tautology; it entails no content elements and is incompatible with no state descriptions.) It is evident, then, that the two notions of content with which we began merge into one and are neatly captured by the account given by Carnap and Bar-Hillel.

The foregoing approach immediately yields a quantitative measure of content, or more accurately, a large class of such measures. In his explication of 'degree of confirmation', Carnap has shown how various confirmation functions could be defined in terms of measures assigned to state descriptions. The measure $m(p)$ of any statement p is then taken to be the sum of the measures of the state descriptions in $R(p)$.[19] If each content element is given a *content measure* equal to the measure of the state description it negates, then the content measure $cont(p)$ will be the measure of the range of $\sim p$. If the measure of p is taken as the initial probability (null confirmation) $c_0(p)$, then we have a suitable sort of inverse relation between initial probability and content measure:[20]

$$cont(p) = m(\sim p) = 1 - m(p) = 1 - c_0(p).$$

There can be little doubt that the foregoing analysis of content is useful and illuminating, but it cannot be used directly as a basis for an analysis of complete independence of statements.[21] The reason is obvious. Given any two statements p and q, each of them entails $p \vee q$, so if we think of content in terms of entailments, any two statements – with one exception to be noted immediately – will have common content by virtue of a common consequence. The one exception arises when the statements p and q are mutually exhaustive, for in that case $p \vee q$ is L-true and has no content, so it can hardly represent shared content for p and q. For example, consider the statements $(\exists x) Fx$ and $(\exists x) \sim Fx$, which are mutually exhaustive but not mutually exclusive. Content element 8 is the entire content of the former, while content element 1 is the entire content of the latter. This is a case of a pair of statements with mutually exclusive content, i.e., no content in common. These two statements are, of course, far from independent, for there is a full entailment relation between them: the falsity of either entails the truth of the other. Thus, nonoverlapping content is no more indicative of complete independence than is nonoverlapping of range. This is no surprise, for the concepts of content and range are essentially dual to one another, so there is not much hope of getting our relation of complete independence out of one if it does not come out of the other.

There is some danger, I fear, that these results may seem so highly counter-intuitive as to suggest that the whole analysis of content has been fundamentally misguided. A little reflection can, I believe, dispel this initial reaction (if it in fact occurs). If there is a relation of complete independence – i.e., if there are statements between which neither full nor partial entailment holds – then consider a paradigm pair such as Fa and Fc. Obviously they have overlapping content just as they have overlapping ranges; they share content elements 6 $(Fa \vee \sim Fb \vee Fc)$ and 8 $(Fa \vee Fb \vee Fc)$. The fact that these independent statements have common content elements should occasion no surprise; in fact, this is precisely what we should have expected. If each of two independent statements rules out sets of possibilities, some of the possibilities ruled out by one will be excluded by the other as well, and some not. The fact that they are independent is *attested* by the fact that neither all nor none of the possibilities incompatible with the one are excluded by the other as well. If it were otherwise, the statements would surely be far from independent. For example, Fa and Fc each rule out state description 8 $(\sim Fa \,\&\, \sim Fb \,\&\, \sim Fc)$, but Fa rules out state descriptions containing $\sim Fc$ indiscriminately, for it also rules out state description 7 $(\sim Fa \,\&\, \sim Fb \,\&\, Fc)$ which is just like state description 8, except for containing Fc where 8 contains $\sim Fc$. Fc does not rule out state descriptions containing $\sim Fc$ in this indiscriminate manner, though it does rule out state descriptions containing $\sim Fa$ indiscriminately. This situation does not seem counterintuitive after all. It strongly suggests that what we began by regarding as a relation of *having no common content* needs more properly to be explicated as a relation of *independence of content*. Indeed, when we succeed in characterizing the relation of complete independence, in terms of which we shall be able to define partial entailment, a suitable concept of independence of content will emerge automatically.

The discussion of content was undertaken to see whether it would be feasible to explicate partial entailment in terms of possession of common content, and complete independence in terms of lack of common content. The result is essentially the same as was obtained in considering relations between ranges: if partial entailment is defined in terms of overlapping content, then there is no such relation as complete independence, for when the content is nonoverlapping a full entailment relation holds. I do not mean to maintain now, any more than I did in the discussion of overlap of ranges, that it is impossible to provide a stipulative definition of 'partial entailment' according to which it obtains whenever there is overlapping of content. Obviously this can be done. However, the consequence is the same as before. The relation of partial entailment that results from such a definition is radically incongruous with that which arises naturally out of the consideration of the relations to be found in deductive logic. Any heuristic value of the analogy between full entailment and partial entailment *vis-à-vis* the foundations of inductive logic is entirely lost.

V. COMPLETE INDEPENDENCE

I have taken *Fa* and *Fc* as a paradigm of complete independence, but any two distinct atomic statements would serve equally well. Likewise, I have taken *Fa*&*Fb* and *Fb*&*Fc* as a paradigm of partial entailment, and thus, of the absence of complete independence. The difference, as already noted, seems to be the presence of common conjuncts in the case of nonindependence, and the absence of common conjuncts in the case of complete independence. However, this cannot be used as a criterion for distinguishing between the two relations. If p and q are L-equivalent, respectively, to p' and q', then surely p must be completely independent of q if p' is completely independent of q'. Now Fa and Fb are L-equivalent, respectively, to $(Fa \vee Fb) \& (Fa \vee \sim Fb)$ and $(Fa \vee Fb) \& (\sim Fa \vee Fb)$, where the latter pair share a common conjunct and the former pair are, by hypothesis, completely independent. As a matter of fact, any statement is L-equivalent to the conjunction of its content elements, so any two statements that have common content are L-equivalent, respectively, to a pair of statements that share a common conjunct. As we have seen, only L-disjoint statements fail to have common content.

If we impose a condition that statements be reduced to some sort of minimal form before we apply the criterion of shared conjuncts, a different difficulty will arise. $Fa \vee Fb$ and $Fb \vee Fc$ are not completely independent of each other, but they do not share a common conjunct. They do, however, share a common atomic part. Thus, we might make the absence of common nonvacuous atomic parts the defining condition for complete atomic independence. More precisely,

> P is *completely atomically independent* of $q =_{\mathrm{df}} p$ and q are L-equivalent respectively to two molecular (quantifier free) statements p' and q' which share no common atomic parts.

This definition will be more useful if we can provide an effective procedure for finding out whether the statements p' and q' exist. This can be done through the use of normal forms. By application of a set of rules, which I shall sketch, any statement of L_N can be reduced to minimal disjunctive normal form.[22] To reduce a statement p to minimal form, one begins by replacing universal and existential quantifiers by conjunctions and disjunctions respectively, thereby reducing the statement to a molecular form. Truth-functional connectives other than negation, conjunction, and disjunction are eliminated according to standard definitions. Rules are provided for the elimination of redundant parts of disjunctions and conjunctions. DeMorgan's rules can be used to break up negations of large scope, and double negation is used to eliminate excess negations. The distribution rule breaks up conjunctions that contain disjunctions as conjuncts. The result will be a statement p' that is L-equivalent to p, and which contains no vacuous atomic parts. It will have the form of a basic statement standing alone, a conjunction of basic statements in which no atomic statement occurs more than once, or a disjunction of such conjunctions of basic statements. Two statements p and

q will be completely atomically independent iff their minimal disjunctive normal forms have no atomic parts in common.

I am confident that complete atomic independence provides at least a sufficient condition of the kind of complete independence we seek. I am less sure that it is an appropriate necessary condition, for another definition of complete independence has a degree of plausibility. For languages of the type we are considering, complete independence would seem to be complete truth-functional irrelevance. We want to find a way of saying that the statement p makes no difference to the truth-functional character of q; i.e., the truth or falsity of p has no bearing on the truth-value pattern of q. This condition obtains, I would suggest, if, in a truth-table adequate to represent the truth patterns of the statements involved, the proportion of true cases for q, given that p is true, is equal to the proportion of true cases for q, given that p is false. This condition is equivalent to the condition, cited by Kemeny and Oppenheim, that there be a zero correlation between the state descriptions in which p holds and the state descriptions in which q holds.[23] Let p and q be neither L-true nor L-false, and let $n(p)$ be the number of cases in which the truth value of p is T. Then,

p is *completely truth-functionally independent* of
$$q =_{\text{df}} n(p \,\&\, q)/n(p) = n({\sim} p \,\&\, q)/n({\sim} p).$$

It follows immediately that, in case of complete truth-functional independence, the proportion of cases of q true given that p is true, equals the proportion of cases of q true regardless of the truth value of p:

$$n(p \,\&\, q)/n(p) = n(q)/n(t),$$

where t is a tautology.[24] Moreover, complete truth-functional independence is, as is to be expected, a symmetrical relation, for from the definiens we get the result that q is completely truth-functionally independent of p; i.e.,

$$n(q \,\&\, p)/n(q) = n({\sim} q \,\&\, p)/n({\sim} q).^{25}$$

Complete truth-functional independence, and its absence, are illustrated in Table I. This definition of complete truth-functional independence translates immediately into a definition in terms of ranges of statements. Simply read '$n(p)$' as the number of state descriptions in the range of p,

TABLE I

p	q	r	p & q	q & r	p ∧ q	q ∧ r	p	p ≡ q
T	T	T	T	T	T	T	T	T
T	T	F	T	F	T	T	T	T
T	F	T	F	F	T	T	T	F
T	F	F	F	F	T	F	T	F
F	T	T	F	T	T	T	F	F
F	T	F	F	F	T	T	F	F
F	F	T	F	F	F	T	F	T
F	F	F	F	F	F	F	F	T

Let t = tautology.

Examples of nonindependence

$$\frac{n[(p \,\&\, q) \,\&\, (q \,\&\, r)]}{n(p \,\&\, q)} = \frac{1}{2} \neq \frac{n[(\sim (p \,\&\, q) \,\&\, (q \,\&\, r)]}{n[(\sim (p \,\&\, q)]} = \frac{1}{6} \neq \frac{n(q \,\&\, r)}{n(t)} = \frac{1}{4}$$

$$\frac{n[(p \vee q) \,\&\, (q \vee r)]}{n(p \vee q)} = \frac{5}{6} \neq \frac{n[\sim (p \vee q) \,\&\, (q \vee r)]}{n[\sim (p \vee q)]} =$$

$$= \frac{1}{2} \neq \frac{n(q \vee r)}{n(t)} = \frac{3}{4}.$$

Examples of independence

$$\frac{n(p \,\&\, q)}{n(p)} = \frac{n(\sim p \,\&\, q)}{n(\sim p)} = \frac{n(q)}{n(t)} = \frac{1}{2}$$

$$\frac{n[p \,\&\, (p \equiv q)]}{n(p)} = \frac{n[\sim p \,\&\, (p \equiv q)]}{n(\sim p)} = \frac{n(p \equiv q)}{n(t)} = \frac{1}{2}.$$

and the definition stands unaltered. I shall return below to the obvious relation between this definition and the confirmation function $c^{\dagger}$.

Complete truth-functional independence is not the same as complete atomic independence. Complete atomic independence implies complete truth-functional independence – this is a direct consequence of our methods for constructing truth-tables. If you make a truth-table for a formula with $n-1$ atomic parts, but with all of the cases for n atomic parts, you simply repeat the truth-value pattern for that formula, producing the truth-value pattern once for the value T of the extraneous atom, and reproducing it for the value F of the extraneous atom. If the truth-table contains more than one extraneous atomic formula, the truth pattern for the original formula will be reproduced for each combination of truth-values for the extraneous atoms. Such repetitions cannot change the ratios involved in the definition of complete truth-functional independence, for the net effect is simply to multiply numerator and denominator by the same integer. Thus, given two formulas p and q with no atomic parts in common, the proportion of true cases of either is not a function of the truth-value of the other. To show that complete truth-functional independence does not imply complete atomic independence, we can provide a counterexample. It is seen immediately, by counting cases, that Fa and $Fa \equiv Fb$ are completely truth-functionally independent (see Table I). It is evident, however, that $Fa \equiv Fb$ is not L-equivalent to any statement not containing Fa. The minimal disjunctive normal form of $Fa \equiv Fb$ is $Fa \,\&\, Fb \vee {\sim}Fa \,\&\, {\sim}Fb$, while the minimal disjunctive normal form of Fa is itself.

Do we want to say that Fa partially entails $Fa \equiv Fb$? The reasons seem to me to favor denying it. Without regard to the truth-value of Fa, $Fa \equiv Fb$ is true in half of the possible cases; given that Fa is true, it is still true in half of the remaining possible cases. To be sure, $Fa \equiv Fb$ is L-equivalent to $(Fa \supset Fb)$, $({\sim}Fa \supset {\sim}Fb)$, neither of whose conjuncts is completely independent of Fa. This does not pose any great obstacle to independence, however, for we are basically concerned with inductive relevance, and it is well-known that evidence can be relevant to each of two statements without being relevant to their conjunction.[26]

VI. PARTIAL ENTAILMENT

The main reason for concern over complete independence is to use it in clarifying the concept of partial entailment. Since partial entailment, as we shall construe it, is a relevance relation, we shall have to make allowance for positive and negative relevance. In case p is positively relevant to q we shall say that p partially entails q; if p is negatively relevant to q we shall say that p partially entails ${\sim}q$. The general idea is that a partial entailment (positive or negative) shall obtain from p to q iff neither a full (deductive) entailment nor complete truth-functional independence obtains. In this discussion we shall continue to restrict p and q to statements that are neither L-true nor L-false.[27] The following definitions seem reasonable:

p *partially entails* $q =_{\text{df}} n(p \,\&\, q)/n(p) > n(\sim p \,\&\, q)/n(\sim p)$, and p does not entail q.

p *partially entails* $\sim q =_{\text{df}} n(p \,\&\, q)/n(p) < n(\sim p \,\&\, q)/n(\sim p)$, and p does not entail $\sim q$.

We shall say informally that there is a partial entailment from p to q if either of the foregoing relations holds. Given these definitions, it will be useful to look once more at the comparison between *logical relations* and *relations of ranges* given in a preliminary way in Figure 1; Figure 2 now provides a more accurate representation. This diagram shows clearly how inclusion and exclusion of ranges constitute entailment relations, while overlap of ranges includes both types of partial entailment as well as complete independence. It should be explicitly noted that complete independence is not necessarily represented diagrammatically by the

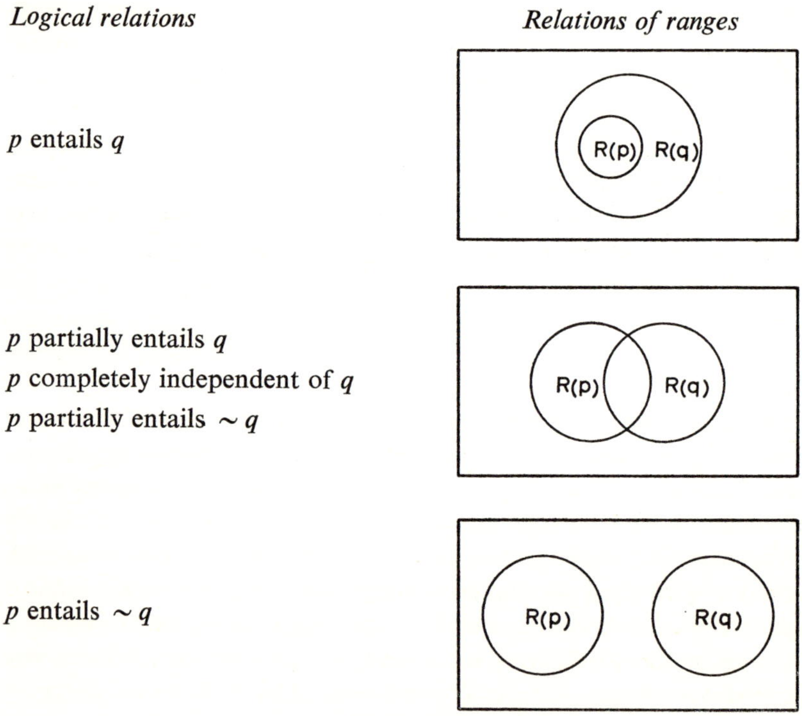

Fig. 2. Two congruous scales.

picture in which $r(p)$ covers half of $r(q)$ and conversely. Complete independence is more complicated; it obtains when the proportion of $r(p)$ overlapped by $r(q)$ equals the proportion of the universe $r(t)$ occupied by $r(q)$. The universe is shown in the diagrams of Figure 2 because of this relation.

Now that the relations of partial entailment and complete independence have been defined, let us look at some of the typical kinds of problems that concern inductive logic. The difficulty that immediately arises is that there are pairs of statements which are completely independent on our definition of complete truth-functional independence (or on the definition of complete atomic independence of that matter) between which inductive logicians have claimed that a relation of partial entailment holds. Consider, for example, the classic set-up for drawing balls from an urn without replacement. Each draw reveals the color of a different ball. Suppose we draw two balls and find that each is red, and we wonder whether the next ball to be drawn will also be red. Letting F be the property red, our 'evidence' is $Fa \& Fb$. We are entertaining the 'hypothesis' Fc. These statements are completely independent on either definition: they have no nonvacuous atomic parts in common, and they are truth-functionally irrelevant to one another. Or, consider the same situation except that each ball is replaced after it is drawn. Different draws need not involve distinct balls. If our individual constants denote events, such as draws from the urn, everything is the same as before. This is the way it would be handled in the Carnapian framework. But even if individual constants must designate physical things so that the 'evidence' and 'hypothesis' may be about the same things, they will still be completely independent.

These two situations give us typical inductive examples. As a matter of fact, the fundamental inductive problem concerns inference from the past to the future, or more generally, from the observed to the unobserved. Past events are distinct from future events; observed events are distinct from unobserved events. Therefore, statements exclusively about past events are completely atomically (and *a fortiori* truth-functionally) independent of statements exclusively about future events, and statements strictly about observed events are completely independent of statements strictly about unobserved events. Hume made this point quite explicitly, and we see that he was right on either definition of complete independence.

There are, of course, inductive examples in which evidence and hypothesis are not completely independent. For instance, going back to the example of draws without replacement, the hypothesis that all of the balls in the urn are red is *not* completely independent of the evidence that the two balls so far drawn are red. There is, as we noted at the outset, a very reasonable sense in which this evidence partially entails this hypothesis. Peirce and Carnap both saw, however, that one can separate 'tested parts' from 'untested parts' of the hypothesis; for instance, if the urn contains only three balls, $Fa \& Fb$ is a tested part of $(x)\,Fx$, while Fc is an untested part. An untested part is a part which is completely independent of the evidence; indeed, it seems natural to define 'untested part' in terms of complete independence:

> h' is an *untested part* (with respect to e) of $h =_{\mathrm{df}} h$ entails h', and h' is completely independent of e.

You may choose either complete atomic independence or complete truth-functional independence; the problem that remains for inductive logic will be equally severe either way.

VII. INDUCTIVE RELEVANCE

Both Peirce and Carnap recognized the nature of the problem involved in the relation between the evidence and an untested part of the hypothesis.[28] If we construe 'partial entailment' and 'complete independence' in accord with the foregoing definitions, and if we identify inductive irrelevance with the absence of full or partial entailment relations (i.e., with complete independence), this is tantamount to the choice of $c^{\dagger}$ as a confirmation function. It would preclude 'learning from experience', for it makes the past inductively irrelevant to the future, the observed inductively irrelevant to the unobserved, and the 'evidence' inductively irrelevant to the untested part of the hypothesis. Peirce reacted to this problem by rejecting the notion that inductive relevance is any kind of logical relation, but Carnap proposed different confirmation functions that do not rule out learning from experience. Carnap does not explicitly recognize the relation of complete truth-functional independence, and he does not see partial entailment as a relation standing somewhere between complete independence and full entailment. Rather, he identifies partial entailment with overlapping ranges. On his conception full or partial entailment always obtains, so inductive irrelevance, if there is to be such a relation, cannot be identified with the absence of full and partial entailments. Carnap does, however, introduce a relation of inductive relevance based upon his confirmation functions,[29] and in the case of $c^\dagger$, inductive irrelevance coincides with our complete truth-functional independence. For other confirmation functions, however, the extension of the relation of inductive irrelevance is different and no longer coincides with complete truth-functional independence. Thus, for example, e may be irrelevant to h on $c^\dagger$, but relevant to h on c^*. Confirmation functions other than $c^\dagger$ provide for the possibility of learning from experience. However, a price is paid for this benefit. On $c^\dagger$, inductive irrelevance is identified straightforwardly with a logical irrelevance relation, but other confirmation functions yield inductive irrelevance relations that have no simple coordination with any logical irrelevance relation. In order to achieve fuller clarity on these matters, it will be useful to examine the concept of inductive relevance in some detail.

One of Carnap's chief contributions to inductive logic, in my opinion, is his careful attention to the concept of inductive relevance, and his emphasis upon the importance of that relation. Not only does he devote a major chapter of *Logical Foundations of Probability* to the analysis of relevance, but also, in his 'Preface to the Second Edition', he emphasizes the importance of distinguishing clearly between two sets of concepts, namely, 'concepts of *firmness*' and 'concepts of *increase of firmness*'.[30] The concepts of firmness have to do with the degree of confirmation of a hypothesis on given evidence; the concepts of increase of firmness have to do with the *change in* degree of confirmation as a result of additional evidence. Change in degree of confirmation is, of course, an *inductive relevance* concept. If h has a certain degree of confirmation $c(h, i)$ on prior evidence i (in case there is no prior evidence, let i be the tautological evidence t), then e is positively relevant to h provided $c(h, i \& e) > c(h, i)$, *negatively relevant* if $c(h, i \& e) < c(h, i)$, and *irrelevant* if $c(h, i \& e) = c(h, i)$. We can define a relevance measure, degree of increase of degree of confirmation, as follows:

$$d(h, i \& e) = c(h, i \& e) - c(h, i).$$

This quantity is obviously negative when e is negatively relevant, and zero when e is irrelevant. When i is the tautological evidence, we may write

$$d(h, e) = c(h, e) - c(h, t).$$

The point on which confusion is most apt to arise is in a failure to distinguish carefully between zero degree of confirmation and zero degree of increase of confirmation, especially if the prior evidence is tautological, i.e., between $c(h, e)=0$ and $d(h, e)=0$. These are clearly not the same. In the finite languages L_N, h has zero confirmation on e iff e is not L-false and e entails $\sim h$. Moreover, h has zero confirmation on t iff h is L-false. Therefore, $c(h, e)=d(h, e)=0$ iff e is not L-false and h is L-false. In all of the interesting cases, namely those in which neither e nor h is L-false, $c(h, e)=0$ and $d(h, e)=0$ are mutually incompatible.

Since Carnap has identified partial entailment with degree of confirmation, e partially entails h if e is not L-false, and e and h are not L-exclusive, and e does not fully entail h (in the finite languages L_N). As emphasized above, it follows that e partially entails h whenever e and h are logically independent. Thus, Carnap makes partial entailment a 'concept of firmness' and not a 'concept of increase of firmness'. In general, e partially entails h even if e is inductively irrelevant to h. The concept of 'learning from experience', in sharp contrast, is a relevance concept or a 'concept of increase of firmness'; it concerns the possibility of evidence raising or lowering probabilities. Therefore, if we take partial entailment as Carnap does, there is no guarantee that learning from experience is possible whenever we have partial entailments; quite the contrary. The confirmation function $c^\dagger$ provides an 'inductive logic' with all the partial entailments Carnap demands, but in which learning from experience is impossible.

One of the many ways of looking at the fundamental problem of inductive logic is to view it as the problem of finding some relation of relevance between past and future, between observed and unobserved, which enables learning from experience to occur.[31] Inductive logic without appropriate relevance relations between past and future is useless. We can, to be sure, list the logical possibilities that may be realized in the future, and we may even assign measures to these possibilities on some *a priori* basis. If, however, experience and observational evidence cannot *alter* the status of these abstract logical possibilities by providing inductive grounds for expecting some more and others less, then inductive logic provides no means by which to overcome our complete ignorance of the future. Regardless of what we have experienced and observed, we would always face the future as nothing more than a set of abstract logical possibilities.

If one maintains that a logical relation of partial entailment holds between evidence and hypothesis, and that this is the foundation of inductive logic, it seems reasonable to expect this relation to be a relevance relation. Carnap's relation of partial entailment simply is not a relevance relation, and it does not pretend to be. As we have seen, Carnap introduces a relevance relation which, in the case of $c^{\dagger}$, coincides with our relation of complete truth-functional independence, but fails to coincide for other confirmation functions. Carnap's approach takes the nonrelevance relation of overlapping ranges as the basis for the concept of partial entailment and degree of confirmation. The particular confirmation function depends upon the way in which measures of overlapping are assigned. Inductive relevance is defined on the basis of degree of confirmation, so relevance relations can be changed by changing the measures assigned to the overlap of ranges. This means that inductive irrelevance, on Carnap's approach, is not tied directly to any logical irrelevance relation that is easily recognizable as such. The fact that inductive relevance and irrelevance are defined in terms of measures of overlap of ranges, and not on the basis of a well-defined logical relation like our complete truth-functional independence, fosters the idea that adjustments in the weights assigned to state descriptions constitutes a reasonable way to change the extension of the inductive relevance relation, thus providing for inductive relevance and 'learning from experience' in those cases in which we feel we should have it in inductive logic.

I am arguing, however, that the search for a *logical* relation upon which to ground inductive logic should proceed from an examination of the familiar relations of deductive logic – at least it should if the analogy between the logical relations of inductive and deductive logic is to have any value for our understanding of the foundations of inductive logic. Our examination of these relations has revealed a relation of maximal relevance (full entailment), a relation of irrelevance (complete truth-functional independence), and an intermediate relation of some sort of logical relevance (partial entailment). If we identify inductive relevance with partial entailment of this type (as opposed to the nonrelevance relation of overlapping ranges), and inductive irrelevance with complete truth-functional independence, then we have a strong analogy between the logical relations of deduction and those of induction. Unfortunately, these identifications make inductive logic entirely sterile by yielding irrelevance in just those cases in which relevance is needed if learning from experience is to be possible. If, on the other hand, we tinker with measures of overlapping, so as to introduce relevance relations that can serve as a basis for learning from experience, then we destroy the coincidence between inductive irrelevance and its straightforward deductive counterpart (complete truth-functional independence). Thus, for confirmation functions other than $c^{\dagger}$, no benefit accrues from the analogy between inductive irrelevance and logical irrelevance, and similarly, none accrues from any analogy between inductive relevance and any logical relevance relation.

We have already seen that two distinct concepts of partial entailment seem to be involved in the doctrine that inductive logic is grounded in a logical relation, and these two concepts are radically incongruous with one another. They are incongruous because one is a relevance relation and the other is not. The relevance relation yields a strong analogy between inductive and deductive logic, but it fails to provide the kind of inductive relevance that is needed for an inductive logic. When we use the nonrelevance concept, it is possible to endow inductive logic with more satisfying relevance relations, but we lose the heuristic value of the analogy. This is not to say we cannot define inductive relevance relations which allow for learning from experience; Carnap has done it. But the burden has been shifted; no longer can we argue that a strong analogy between relations of full and partial entailment makes the fundamental character of inductive logical relations intelligible; instead we must argue that the existence of intelligible inductive logical relations demonstrates an analogy between inductive and deductive logical relations. The heuristic value of the analogy is similarly destroyed when one departs from the clear concept of complete truth-functional independence in order to arrive at a more satisfactory relation of inductive irrelevance. There seems to be no unequivocal set of concepts that preserves the analogy *and* provides the relevance relations needed for inductive logic.

VIII. LOGICAL VS. FACTUAL RELATIONS

The foregoing analysis of the relevance and irrelevance relations to be found in deductive logic, and their comparison with those needed in inductive logic, seem to me to strongly suggest that the conception of inductive logic as based upon a logical relation is fundamentally misconceived. There are two distinct kinds of relevance relations, logical and factual, and I am inclined to believe that although deductive logic requires and exploits logical relevance relations, induction is involved with factual relevance relations instead.

When I suggested above that any two distinct atomic statements constitute a paradigm pair of completely independent statements, I was, of course, referring to complete independence of some logical sort. I meant to make no suggestion about their factual independence. To take the most extreme case, let F and G be two distinct primitive predicates, and let a and b designate two distinct individuals. Fa and Gb are completely independent in any logical sense, but there may be a factual relation between them. It may be, for example, that $(x)\,(y)\,[Fx \,\&\, Rxy \supset Gy]$ is a law of nature, and that Rab is true as a matter of contingent fact. Under these circumstances there is a strong dependency between Fa and Gb – i.e., Fa materially implies Gb – but this is not based upon any logical relation between them. Similarly, some strong statistical relation might obtain between F's and G's which would make Fa constitute excellent inductive grounds for asserting or betting on Gb, but again, it need not be any sort of logical relation.

Let me make it entirely clear that I am not asserting that the entailment, partial entailment, and independence relations I have discussed are the only syntactical or semantical relations that can hold among the sentences of our languages L_N. For example, the primitive symbols of a language can be given an alphabetic order in terms of which the well-formed formulas of the language can be alphabetized. Sentences of the language can then stand in syntactical relations of before or after in the alphabetic order. Likewise, we could establish equivalence classes of L-equivalent sentences (such an equivalence class seems to represent what has traditionally been known as a proposition), and these equivalence classes could be ordered in terms of the length of their alphabetically first members. Moreover, it is possible to define 'degree of confirmation' in such a way that true degree of confirmation statements are analytic; in this sense, degree of confirmation does constitute a logical relation. I do not intend to claim that there are no such logical relations, but only that they are not germane to inductive logic.

Consider the term 'entailment'. If we take it as an uncommitted symbol, we can assign it any meaning we wish, e.g., parenthood. If we decide to let it stand for some logical – i.e., syntactical or semantical – relation, we can define it so that *p* is said to entail *q* iff *p* is alphabetically prior to *q*. If we want it to represent a full-fledged semantical relation, we could stipulate that *p* entails *q* iff *q* L-implies *p*. However, all such definitions are obviously grossly inadequate as explications of the notion of entailment with which we are normally concerned. Entailment is a semantical relation that has to do with truth-values. To use the word 'entailment' for some completely nonlinguistic relationship, such as the relation of parenthood, is blatantly inadequate as an explication. It is likewise unsatisfactory to use this term to refer to a purely syntactical relation that has nothing to do with truth-values, e.g., the relation of alphabetic precedence. Moreover, it is unsound to use the word to designate the semantic relation which is the converse of L-implication. 'Entailment' must refer to a relation between statements that is necessarily truth-preserving, that is to say, a relation such that it is impossible to have *p* be true and *q* false if *p* entails *q*. Happily, we find it possible to define 'entailment' in terms of L-implication in such a way that the semantic rules of the language guarantee that the relation so designated will have the desired characteristics.[32]

What I have said about entailment seems obvious enough, but I have spelled it out in laborious detail in order to have a basis for comparison with degree of confirmation. 'Degree of confirmation', as well as a host of other related terms such as 'inductive support', 'inductive relevance', and even 'partial entailment', are not completely uncommitted symbols. They do not call for stipulative definitions; instead, they demand explication. The aspect of the explication that requires closest attention, it seems to me, is its involvement with the truth-values of statements. To define 'degree of confirmation' in such a way that it has nothing to do with the truth-values of the statements bearing the relation of confirmation misses the point in just the same way as would an explication of 'entailment' that referred only to the alphabetic ordering of sentences. In logic, whether it be deductive or inductive, we are concerned with inference – that is, we might say, with the truth-values of some statements given truth-values of others. In deductive logic we have a logical relation of entailment which guarantees a truth-value of one statement, given a truth-value of another, and this guarantee depends only upon the semantical and syntactical rules of the language in which the statements occur. Another relation which has a similar character, even though it is a factual relation rather than a logical relation, is material implication. If p materially implies q, but does not entail q, then q is true if p is, but we cannot certify that this relation obtains between two statements on the basis of semantical and syntactical relations alone.

In addition to the logical relation of entailment, we have found a logical relation of partial entailment as well. Whether this relation holds between two statements p and q depends only upon their syntactical and semantical features. When it does obtain, there is, as we have seen, an alteration of the truth-value pattern of q depending upon whether p is true or false. The truth-value pattern of q is a function of the truth-value of p.

Our main question is whether a logical relation of confirmation can provide a foundation for inductive logic, that is, a basis for inductive inference. Without inductive inference we remain totally ignorant of the future. We can view it only as a set of abstract logical possibilities about which we might guess blindly or simply confess ignorance. The function of inductive logic is to provide a basis for inference which is some improvement over total ignorance. If we introduce the confirmation function $c^{\dagger}$ we find no improvement in our inductive situation at all. Whatever evidence we have is irrelevant to the future. With this confirmation function, inductive irrelevance coincides with complete truth-functional independence. Degree of confirmation, using $c^{\dagger}$, does have some direct relation to truth-values, but it is no good as a foundation for inductive inference. Our real need, as we have seen, is not so much for a logical confirmation function as for a logical inductive relevance relation. In order to provide a suitable foundation for inductive logic, we need an inductive relevance relation that yields relevance even for statements that are completely truth-functionally independent. And this relevance relation must have some bearing upon inference.

The desideratum for inductive logic is to enable us to make inferences about the future that are some improvement upon the sheer blind guessing for which we would have to settle in the absence of inductive logic. This suggests – to me, irresistably – that we are looking for a relation upon which to base inductive inference that will enable us to be right more often than could be expected by blind guessing. In other words, the relation upon which inductive inference is to be based is designed to yield true results more frequently when we take account of relevant evidence than we get by ignoring it. For example, we should be right more often in inferring or betting on *Fc* in the presence of *Fa*&*Fb* than we would be in inferring or betting on *Fc* either in the absence of relevant evidence or in the presence of negative evidence such as ~*Fa*& ~*Fb*.

Unfortunately there is no known logical relation for which any such claim can be substantiated. While it is possible to define a logical (i.e., syntactical or semantical) relation between evidence and hypothesis which can be called 'degree of confirmation', this relation has no useful bearing upon truth-values of hypotheses in relation to evidence. If there are logical dependencies among truth-values, any satisfactory concept of degree of confirmation will reflect them. For example if p entails q, the degree of confirmation of q on p must be one, and if p entails $\sim q$, the degree of confirmation of q on p must be zero. Moreover, if there is a positive correlation between the truth-cases for p and the truth-cases for q – i.e., a positive correlation between the state descriptions in which p holds and those in which q holds – we may expect this to be reflected as positive inductive relevance.[33] When, however, there is no functional relation between the truth-value pattern of p and the truth-value pattern of q, the procedure usually followed is to assign different weights to different total states of the universe (state descriptions). For example, Carnap's well-known confirmation function c^* results for L_3 when we assign the value 1/4 to state descriptions 1 and 8, and the value 1/12 to the remaining ones.[34] If we find, by observation, that *Fa* holds, this eliminates low-valued state descriptions in which *Fc* holds, and leaves the relatively higher valued one, whereas, ~*Fa* eliminates the relatively high-valued one and leaves the low-valued state descriptions in which *Fc* holds. This is the source of the positive relevance of *Fa* to *Fc*. As we have noted repeatedly, if we assign equal weights to all of the state descriptions, namely, 1/8 to each, *Fa* is irrelevant to *Fc*. Of course, it is also possible to give state descriptions 1 and 8 a weight below that of the others; for instance, we can assign a weight of 1/14 to each of them, and let 2–7 each have the value 1/7. The net result will be to make *Fa* negatively relevant to *Fc*, while making ∼*Fa* positively relevant to *Fc*. For any two statements that are completely truth-functionally independent of each other, positive inductive relevance, negative inductive relevance, and inductive irrelevance are *determined entirely* by the *a priori* assignment of weights to the possible states of the universe.

If we knew in advance what state of the universe is the actual one – that is, if we knew which state description correctly characterizes the observed and unobserved portions of the universe – we would have no need for inductive logic. Our problem arises because we have no such knowledge. If, to borrow Peirce's apt phrase, universes were 'as plentiful as blackberries' we might try counting to see how often a particular state description turns out to be correct. As a matter of fact, universes are available in large numbers, for when we speak of a universe it can be a small thing such as an urn with some balls in it or a few tosses of a die. It need not be the entire physical world. A universe is whatever can be completely described by the particular language we are using; the language we have used as an example can describe three things with respect to one property. If universes containing three things characterized by the presence or absence of one property exemplify state descriptions 1 and 8 more frequently than the other state descriptions, the kind of relevance relations established by c^* will yield increased truth-frequency for inductive inferences. If the situation is reversed, with 1 and 8 being realized relatively infrequently, inference based on c^* will lead to decreased truth-frequency. The essential point is simply that no syntactical or semantical features of state descriptions can be shown to have any relation whatever to the frequency with which they are realized in actual universes. Thus, there is no *logical* relation among state descriptions, and consequently no logical relation between evidence and hypothesis, that has any bearing upon the truth-frequencies that will result from the choice of any particular confirmation function. We can define an unlimited supply of confirmation functions, each of which gives us different relevance relations, but we cannot establish anything about the truth-frequencies any of them will yield if used as a basis for inductive inference. They are all as irrelevant to the aims of inductive inference as is the definition of 'entailment' in terms of alphabetic precedence to the aims of deductive logic.

Given a particular definition of 'degree of confirmation', it is impossible to show that one will be right more often by inferring or betting on highly confirmed hypotheses than on those with lower degrees of confirmation. If certain *factual* relations obtain – e.g., if F is a property that usually occurs uniformly in universes of the type we are considering – then utilization of degrees of confirmation based upon a confirmation function like c^* will have some benefit in inductive inference. The benefit derives from a *factual* relation, however, not from a logical one. Similarly, if p materially implies q without entailing q, inferring q from p will be successful, but not because of any logical relation.

I am fully aware that strenuous objection will be raised against the view I have been proposing. Given the truth of e, we will be told, the logical relation of confirmation cannot guarantee either a truth-value of h or a truth-frequency for hypotheses which, like h, have a certain degree of confirmation with respect to e, but it can tell us what is reasonable to believe about the truth-value of h. I must confess that I find it impossible to understand the force of the concept of rationality employed in this way. If adoption of an inductive logic does not give promise of greater success than sheer guessing, what is reasonable about it? If a particular inductive logic, based upon some particular definition of 'degree of confirmation', is supposed to give greater promise of success, in what does that promise lie?[35] There are, as remarked above, two kinds of relevance relations, deductive and inductive. Deductive relevance is to be found in relations of entailment and partial entailment; these are logical (syntactical and semantical) relations. Inductive relevance in contrast, seems to be fundamentally a relation of statistical relevance. Statistical correlation is an example of an inductive relevance relation, and so is material implication as a special limiting case of statistical correlation. Unlike deductive relevance relations, these relations cannot be established on the basis of semantical and syntactical considerations alone. They are factual relations. I am convinced that inductive logicians have been interested in confirmation because they believe that the logical relation of degree of confirmation mirrors the factual statistical relevance relations that actually obtain in the world. Such an assumption is entirely gratuitous; there is no way of establishing the correspondence of the logical with the factual relations. The search for logical surrogates for these factual relations appears to be futile.

IX. CONCLUSION

The idea that inductive logic is based upon a *logical* relation between evidence and hypothesis – a relation identified as partial entailment – seems to me to rest upon an illusion. The illusion is fostered by a failure to make several important distinctions. First, we must distinguish the logical relation between evidence and hypothesis from the logical relation between evidence and an *untested part* of a hypothesis. Sometimes there is, in a plausible sense, a logical relation of partial entailment between evidence and hypothesis. This does not help as a foundation of inductive logic, however, for the relation does not hold between the evidence and any untested part of the hypothesis.

Second, we must distinguish statements whose content is mutually *exclusive* from those whose content is mutually *independent*. It is tempting to think preanalytically that 'p partially entails q' means 'p entails part of what q asserts', and that 'p does not (at least) partially entail q' means 'p does not entail part of what q asserts'. Analysis shows, however, that this way of speaking can be misleading unless we understand that not entailing part of what q asserts means that the content is independent of the content of q, *not* that it is exclusive of the content of q.

Third, and most important, we must distinguish clearly between the relation of nonzero confirmation and that of inductive relevance. We must be careful which of these is made to correspond to partial entailment. It is all too easy to fasten upon the relation, *e does not entail* $\sim h$, and then to slip unnoticed into the supposition that it is a relation of positive relevance of e to h.

Although all of these distinctions are elementary and obvious, I am strongly inclined to suspect that they have not always been kept clear. For this reason I believe that the concept of partial entailment has been misleading in inductive logic. No doubt concepts like partial entailment, content, and independence *can* be defined clearly, and *can* be used in ways that lead to no confusion, but I do not think this has happened in the tradition of the logical theory of probability. Indeed, I suspect that the notion of partial entailment has been a source of confusion so profound that it has led inductive logicians to mistake a factual relation for a logical one.

The spirit of this paper is essentially Humean. I have tried to take seriously, and follow out in detail, Hume's dictum that the future is logically independent of the past. If there are any inductive relevance relations, I have maintained, they must be factual rather than logical. There are no inductive or deductive *logical* relations by which the Humean independence can be circumvented. The arguments I have offered, if sound, point to the futility of approaching inductive logic via the logical concept of probability or degree of confirmation.

My sole aim has been to argue against one approach without attempting to suggest or support any alternative. If I am correct in claiming that inductive inference depends upon factual, not logical, relevance relations, the obvious alternative is not attractive. Indeed, we find ourselves immediately confronting another of Hume's major claims, for the factual relations upon which inductive inference seems to be founded apparently cannot be established except by induction. A vicious circle threatens. The escape from this dilemma still constitutes a most serious philosophical challenge.

The foregoing attack upon the logical concept of probability may seem like an inappropriate way to pay homage to an admired friend who is one of the ablest champions of the logical conception, but I am not so immodest as to suppose that my arguments are faultless and definitive. Moreover, Hempel has always prized philosophical clarity above adherence to doctrine. If my discussion can add any clarity at all to the important concepts that provide the bases of inductive logic, I shall have paid him the highest tribute of which I am capable.

Indiana University and
University of Pittsburgh

REFERENCES

[1] Carl G. Hempel, 'A Purely Syntactical Definition of Confirmation', *The Journal of Symbolic Logic* **8** (1943), 122–143; 'Studies in the Logic of Confirmation', *Mind* **54** (1945), 1–26; 97–120 (reprinted, with some changes and with a 'Postscript', in Carl G. Hempel, *Aspects of Scientific Explanation*, The Free Press, New York, 1965); Carl G. Hempel and Paul Oppenheim, 'A Definition of "Degree of Confirmation"', *Philosophy of Science* **12** (1945), 98–115; Olaf Helmer and Paul Oppenheim, 'A Syntactical Definition of Probability and Degree of Confirmation', *The Journal of Symbolic Logic* **10** (1945), 25–60.

[2] For citations to some of the more important articles see footnote 1 to the 'Postscript' in *Aspects of Scientific Explanation*, *op. cit.*, p. 47.

[3] See, for example, John R. Wallace, 'Goodman, Logic, Induction', *The Journal of*

Philosophy **63** (1966), 310–328; Marsha P. Hanen, 'Goodman, Wallace, and the Equivalence Condition', *The Journal of Philosophy* **64** (1967), 271–280; Howard Smokler, 'The Equivalence Condition', *American Philosophical Quarterly* **5** (1967), 300–307.

4 'Inductive Inconsistencies', *Synthese* **12** (1960), 439–469; reprinted with slight changes in *Aspects of Scientific Explanation, op. cit.*; and 'Recent Problems of Induction', in *Mind and Cosmos* (ed. by Robert Colodny), University of Pittsburgh Press, Pittsburgh, 1966.

5 See all of the papers in Parts III and IV of *Aspects of Scientific Explanation, op. cit.*, and also 'Deductive-Nomological vs. Statistical Explanation', in *Minnesota Studies in the Philosophy of Science*, vol. III (ed. by Herbert Feigl and Grover Maxwell), University of Minnesota Press, Minneapolis, 1962.

6 See 'Postscript' in *Aspects of Scientific Explanation, op. cit.*, p. 50.

7 See 'On Vindicating Induction', in *Induction: Some Current Issues* (ed. by Henry E. Kyburg, Jr., and Ernest Nagel), Wesleyan University Press, Middletown, Conn., 1963, and in *Philosophy of Science* **30** (1963), 252–261. This article contains a resolution of Goodman's paradox which I still think to be substantially correct.

8 This study was initiated by consideration of Mary Shearer, 'An Investigation of $c^{\dagger}$', M.A. dissertation, Wayne State University, 1966. In this thesis, Mrs. Shearer undertook to demonstrate Carnap's claim that $c^{\dagger}$ does not admit 'learning from experience'. In carrying out her proof she used the notion I have introduced below under the name of 'complete atomic independence', but it turns out that essentially the same proposition can be demonstrated very much more simply by using what I call 'complete truth-functional independence'. This is a strong motive for introducing the latter as a distinct concept of complete independence.

9 This point was set out sketchily in my article 'Carnap's Inductive Logic', *The Journal of Philosophy* **64** (1967), 21.

10 See 'Preface to the Second Edition', *Logical Foundations of Probability*, University of Chicago Press, Chicago, 1950; 2nd ed., 1962.

11 A full account can be found in Rudolf Carnap, *ibid.*, Chap. III. I shall sketch a few of the essential points, but the foregoing source should be consulted for greater precision and detail. In one deliberate departure from Carnap's usage, I shall use the term 'statement' in almost all cases in which he would use the term 'sentence'. I do this to emphasize the fact that we are dealing with interpreted languages and not just uninterpreted strings of symbols. In order to avoid considerable complication of the notation, I frequently use italicized letters and formulas containing italicized letters as names of themselves. I believe the context makes clear in all cases whether such symbols are being mentioned or only used.

12 See definition D 18–4, *ibid.*, pp. 78 ff.

13 *Ibid.*, § 20.

14 *Ibid.*, § 55B.

15 Throughout this discussion I shall make free use of substitution under L-equivalence, for all of the L-concepts are invariant with respect to such replacement, and the logical relations of inductive logic as discussed by Hempel and Carnap are likewise invariant as long as we do not switch from one language L_N to another.

16 Carnap, *Logical Foundations of Probability, op. cit.*, § 55B.

17 I have expressed both of these notions in very elementary terms in my *Logic*, Prentice-Hall, Inc., Englewood Cliffs, 1963, pp. 14–15; 98–100.

18 Yehoshua Bar-Hillel and Rudolf Carnap, 'Semantic Information', *British Journal for the Philosophy of Science* **4** (1953–54), 147–57.

[19] Carnap, *Logical Foundations of Probability*, *op. cit.*, § 55A.
[20] See Karl R. Popper, *The Logic of Scientific Discovery*, Basic Books, New York, 1959, §§ 31–35, and Rudolf Carnap, 'Probability and Content Measure', in *Mind, Matter, and Method* (ed. by Paul Feyerabend and Grover Maxwell), University of Minnesota Press, Minneapolis, 1966.
[21] Exception could be taken to the Carnap–Bar-Hillel analysis of content on the ground that L-implication is unsatisfactory as an explication of entailment. Anderson and Belnap, for example, have worked extensively on alternative concepts of entailment; see 'Tautological Entailments', *Philosophical Studies* **13** (1962), 9–24. Interesting as these investigations are, they do not seem likely to enable us to salvage the idea that completely independent statements have no common content, for the chief obstacle is the logical principle of addition (p entails $p \vee q$) which holds for the stronger entailment relations as well as for Carnap's L-implication. My own feeble efforts to cook up an entailment relation that would not have this property produced nothing plausible. I am grateful to Professor Alan Anderson for pointing out a severe difficulty in one proposal.
[22] See Carnap, *Logical Foundations of Probability*, *op. cit.*, § 21C, for a full and precise statement of the rules.
[23] John G. Kemeny and Paul Oppenheim, 'Degree of Factual Support', *Philosophy of Science* **19** (1952), 307–324. See especially CA11 and Theorem 2. Letting t be a tautology, $n(t)$ becomes the total number of state descriptions. Kemeny and Oppenheim show that zero correlation obtains iff

$$n(p)\,n(q) = n(p \,\&\, q)\,n(t),$$

which is shown in footnote 24 to be equivalent to our definition.
[24] Given that

$$\frac{n(p \,\&\, q)}{n(p)} = \frac{n(\sim p \,\&\, q)}{n(\sim p)}$$

we have

$$n(p \,\&\, q)\,n(\sim p) = n(\sim p \,\&\, q)\,n(p).$$

Adding $n(p \,\&\, q)\,n(p)$ to both sides yields

$$n(p \,\&\, q)\,n(p) + n(p \,\&\, q)\,n(\sim p) = n(p \,\&\, q)\,n(p) + n(\sim p \,\&\, q)\,n(p)$$

or

$$n(p \,\&\, q)\,[n(p) + n(\sim p)] = n(p)\,[n(p \,\&\, q) + n(\sim p \,\&\, q)].$$

Since $n(t) = n(p) + n(\sim p)$ and $n(q) = n(p \,\&\, q) + n(\sim p \,\&\, q)$,

$$n(p \,\&\, q)\,n(t) = n(p)\,n(q)$$

or, dividing

$$\frac{n(p \,\&\, q)}{n(p)} = \frac{n(q)}{n(t)}.$$

[25] This fact is proved as follows. Assume

$$\frac{n(p \,\&\, q)}{n(p)} = \frac{n(\sim p \,\&\, q)}{n(\sim p)}.$$

Since $n(p) = n(p \& q) + n(p \& \sim q)$ and $n(\sim p) = n(\sim p \& q) + n(\sim p \& \sim q)$,

$$\frac{n(p \,\&\, q)}{n(p \,\&\, q) + n(p \,\&\, \sim q)} = \frac{n(\sim p \,\&\, q)}{n(\sim p \,\&\, q) + n(\sim p \,\&\, \sim q)}.$$

By cross-multiplication,

$$n(p \,\&\, q)\,[n(\sim p \,\&\, q) + n(\sim p \,\&\, \sim q)] = n(\sim p \,\&\, q)\,[n(p \,\&\, q) + n(p \,\&\, \sim q)]$$

or

$$n(p \,\&\, q)\, n(\sim p \,\&\, q) + n(p \,\&\, q)\, n(\sim p \,\&\, \sim q) = n(\sim p \,\&\, q)\, n(p \,\&\, q) + n(\sim p \,\&\, q)\, n(p \,\&\, \sim q).$$

Subtracting $n(p \,\&\, q)\, n(\sim p \& q)$ from both sides yields

$$n(p \,\&\, q)\, n(\sim p \,\&\, \sim q) = n(\sim p \,\&\, q)\, n(p \,\&\, \sim q)$$

which is equivalent to

$$n(q \,\& p)\, n(\sim q \,\&\, \sim p) = n(\sim q \,\& p)\, n(q \,\&\, \sim p)$$

by commutation of conjunction. Adding $n(q \& p)\, n(\sim q \& p)$ to both sides we have

$$n(q \,\& p)\, n(\sim q \,\& p) + n(q \,\& p)\, n(\sim q \,\&\, \sim p) = n(q \,\& p)\, n(\sim q \,\& p) + n(\sim q \,\& p)\, n(q \,\&\, \sim p)$$

or

$$n(q \,\& p)\,[n(\sim q \,\& p) + n(\sim q \,\&\, \sim p)] = n(\sim q \,\& p)\,[n(q \,\& p) + n(q \,\&\, \sim p)].$$

Since $n(\sim q) = n(\sim q \& p) + n(\sim q \& \sim p)$ and $n(q) = n(q \& p) + n(q \& \sim p)$,

$$n(q \,\& p)\, n(\sim q) = n(\sim q \,\& p)\, n(q).$$

Dividing by $n(q)$ and $n(\sim q)$, neither of which is zero,

$$\frac{n(q \,\& p)}{n(q)} = \frac{n(\sim q \,\& p)}{n(\sim q)}.$$

[26] See Carnap, *Logical Foundations of Probability*, *op. cit.*, T 69–1, table 1b, p. 378.
[27] We impose this restriction to prevent the denominators of the fractions in the definitions from being zero. This is no serious restriction, since in inductive logic we are concerned with relations among factual statements.
[28] See Carnap, *Logical Foundations of Probability*, *op. cit.*, §110A, and *The Collected Papers of Charles Sanders Peirce*, 2.744–46.
[29] Chapter VI of *Logical Foundations of Probability* is devoted to the subject of relevance, and it provides considerable illumination for this seriously neglected topic.
[30] Pp. xv–xx.
[31] Though I shall often refer only to inference from past to future, it should be understood that the general problem concerns inference from the observed to the unobserved, regardless of the temporal locations of the events involved.
[32] There are some philosophers, e.g., Anderson and Belnap, *op. cit.*, who would maintain that we still have not found a fully adequate explication of 'entailment', and that a relation must fulfill additional conditions to deserve the name. That does not matter to the point I am making. They would surely agree that the relation must fulfill at least the conditions I have mentioned, and that is all I need to insist upon.
[33] It would be a mistake, I believe, to conclude that this type of positive relevance must be reflected, in turn, in increased truth frequencies, but that is a separate issue right now. The issue at present concerns inductive relevance where no correlations of truth values or state descriptions obtain.

[34] See *Logical Foundations of Probability*, *op. cit.*, 110A, for the definition of c^*. Although Carnap no longer regards c^* as fully adequate, he does maintain that it is a good approximation in many circumstances. In any case, it illustrates nicely the point about relevance under discussion here.

[35] I have discussed the problem of the force of the term 'rational' in inductive logic in several papers, especially in the lead paper and reply to S. Barker and H. Kyburg in 'Symposium on Inductive Evidence', *American Philosophical Quarterly* **2** (1965), 1–16.

WILFRID SELLARS

ARE THERE NON-DEDUCTIVE LOGICS?

I

1. Without attempting to define what is meant by a 'logic', it seems reasonable to say that, however many 'logics' there are, they are 'logics' by virtue of their concern with what makes an argument sound. In the case of 'deductive logic', the concept of a sound argument is that of an argument which is such that if its premisses are true, its conclusion *must* be true. A *good* deductive argument is one which is not only sound (valid) but has true premisses, and hence a true conclusion.

2. An argument purports to be 'deductive' if its conclusion is qualified by the parenthetical adverb 'necessarily', thus:

All men are mortal.	All Texans are rich.
Socrates is a man.	Getty is rich.
So (necessarily) Socrates is mortal.	So (necessarily) Getty is a Texan.

Although both these arguments *purport* to be deductive, the former is correct (valid) and indeed sound, while the second is incorrect (invalid).

3. Notice that the parenthetical comment 'necessarily' does not mean that the conclusion by itself is a necessary truth. The necessity indicated is *relative* necessity; the conclusion (if the argument is sound) is necessary relatively to the truth of the premisses. Of course, the conclusion of an argument may be a *statement of necessity*, thus

> All propositions the denial of which is self-contradictory are necessary.
> The denial of the proposition 'Either snow is white or snow is not white' is self-contradictory.
> So (necessarily) the proposition 'Either snow is white or snow is not white' is necessary.

(Note the occurrence of *both* the parenthetical adverb 'necessarily' and the modal adjective 'necessary'.)

N. Rescher et al. (eds.), Essays in Honor of Carl G. Hempel.

4. In the conclusion of an argument which purports to be deductive

So (necessarily) *p*,

the conclusion *asserts* '*p*' and *signifies* by the parenthetical comment that '*p*' is the conclusion of a deductively good argument, i.e., one which is valid and has true premisses.

5. An argument which purports to be deductive may be incomplete (an 'enthyneme') in the sense that one or more premisses are left to be understood. Thus,

All men are mortal.
So (necessarily) Socrates is mortal.

leaves the premiss 'Socrates is a man' to be understood. If it is supplied, one gets a *complete* argument which not only purports to be deductive, but is actually both valid and good. It is notoriously difficult, in many cases in which one is offered an argument which purports to be deductive, to decide whether the person who offers it is leaving premisses to be understood which, if supplied, would turn it into a valid deductive argument, or whether he thinks, mistakenly, that the argument as it stands is deductively valid. Thus it is often not clear whether a bad argument is intended as complete and should therefore be criticized as invalid, or whether, when understood premisses are supplied, it turns out to be deductively valid, but bad, because these premisses are false.

6. Now, *prima facie*, there are good arguments which, if taken to be complete, are not only *not* deductively valid, but do not even *purport* to be so. Consider, for example

Black clouds are gathering.
So (probably) it will shortly rain.

This argument does not even purport to be deductively valid, as is indicated by the use of the parenthetical comment 'probably' (as contrasted with the 'necessarily' of the arguments we have previously considered). Let us say that it purports to be a good probability argument. Even as a probability argument, however, it is clearly not complete. It is plausible to suggest that we are to supply some such premiss as

When black clouds gather, it usually rains.

When this premiss is made explicit, the argument becomes

> When black clouds gather, it usually rains.
> Black clouds are gathering.
> So (probably) it will shortly rain.

It now strikes us as sound, even though it neither is, nor purports to be, deductively valid. If we suppose that the premisses are true, we are tempted to say that it is not only sound, but good; a good probability argument.

7. Notice that the conclusion does not say

> *It is probable that* it will shortly rain

anymore than the conclusion of 'old mortality' said

> *It is necessary that* Socrates is mortal.

Just as the deductive conclusion *asserted* that Socrates is mortal, and *implied* (or *signified*) that the conclusion was necessary relatively to true premisses; so the conclusion of the present argument *asserts* that it will shortly rain, and *implies* that this conclusion is *probable* relatively to premisses which are true and which satisfy a further condition which can be expressed by saying that the premisses formulate all relevant knowledge. I shall henceforth abbreviate this condition by the letter '*Q*'.

8. In the case of 'old mortality', we pointed out that the adverbial qualifier '(necessarily)' implies that the conclusion is necessary relatively to the premisses (and that the premisses are true). By parity of reasoning we would expect that if our probability argument is good, then the following statement is true.

> 'It will shortly rain' is probable relatively to the premisses. 'When black clouds gather, it usually rains' and 'Black clouds are gathering', which are true and contain all that is known to be relevant.

9. A (complete) deductive argument is good, if its conclusion stands in a certain logical relation to its premisses, and its premisses are true. By parity of reasoning a (complete) probabilistic argument is *good* if its conclusion stands in a certain relation (which it is presumably proper to call a logical relation) to its premisses, and its premisses are true (and satisfy *Q*).

II

10. There are interestingly different varieties of argument having the general form

g & *Q*
So (probably) *p*.

In the first place, there is the distinction between those cases in which we say simply

g & *Q*
So (more probably than not) *p*,

and those in which we say something like

g & *Q*
So (probably to degree .9) *p*.

Abstractly put, we distinguish between a purely *comparative* and a *metrical* concept of probability. If we distinguish, as we are wont, between such species of probability as

The probability of theories
The probability of laws (law-like statements)
The probability of statements of proportion
The probability of singular statements

(a list not intended to be exhaustive), we have in mind such probabilistic arguments as

g & *Q*
So (probably) *T*.

Where '*T*' represents a specific theory;

g & *Q*
So (probably) *LL*,

where '*LL*' represents a law-like statement (i.e., a statement which purports to formulate a scientific law);

g & *Q*
So (probably) n/NC_A is *B*,

where 'n/NC_A is B' says that n/N of a finite class of items of kind A are B; and, to complete our list,

g & Q
So (probably) x_1 is B.

In the last two cases we are often in a position to offer the stronger arguments

g & Q
So (probably to degree .75) n/NC_A is B,

and

g & Q
So (probably to degree .75) x_1 is B.

11. To be confident that we understand the generic concept of probability, i.e., that we understand the nature of probability arguments, and what it is which makes them sound (even though they are not deductively valid), we must examine a variety of these specific forms of argument. To concentrate on one, to take one of them as the 'paradigm', is to run the danger of being misled by its specific character into misconstruing the generic notion of probability, and, therefore, of misconstruing even the 'paradigm'. (For to misconstrue a genus is *ipso facto* to misconstrue all its species.)

III

12. I have argued that if the argument

g & Q
So (probably) p

is to be a good argument, there must be a relation (which can in a suitably broad sense be called 'logical' between 'g & Q' and 'p', such that the truth of the former makes it reasonable to accept the latter, even though the purported deductive argument

g & Q
So (necessarily) p

would be invalid. Now since the relation is to be 'logical' and to be of a kind which makes it reasonable to accept the conclusion if the premisses are true, it seems proper to speak of it as an implication relation, and to characterize it as 'probablistic implication', since it is not the kind of implication which makes deductive arguments valid.

13. This suggests that

g & Q
So (probably) p

is a good probabilistic argument if 'g & Q' is true and 'g & Q' probabilistically implies 'p'.

14. Can we say something more specific about this probabilistic implication? Well what sorts of things can we say about deductive implication which might be suggestive? One familiar line of thought is the following: Consider implication statements of the form

'p and (if p, then q)' (deductively) implies 'q'

and consider the deductively valid arguments it authorizes, schematically,

p and (if p, then q)
So (necessarily) q.

All of these arguments have the property of being truth-preserving, i.e., they are such that if the premiss is true, the conclusion must be true. Now sound probabilistic arguments do not have this property. But perhaps they have *something like* truth-preservingness. Consider,

(3/4C_A is B) & Q
x_1 is C_A
So (probably) x_1 is B.

Arguments of this form are not truth-preserving. It is possible for an argument of this form to be sound, yet have true premisses and a false conclusion.

15. However, let us represent the members of the class C_A by 'a_1', ..., 'a_n' (C_A being a finite class of A's). Consider the class of specific arguments

(3/4 C_A is B) & Q	...	(3/4 C_A is B) & Q
So (probably) A_1 is B		So (probably) A_n is B.

(This class of arguments has n members – for we are concerned with *species* of arguments and not with argument occasions.) We see that 3/4 of these arguments have true conclusions, if the premiss '(3/4 C_A is B) & Q' is true. It seems, therefore, appropriate to say that this class of arguments is 'truth-preserving to degree .75'. By contrast, the class of arguments represented by

All C_A is B
x_i is C_A
So (necessarily) x_i is B

will *all* have true conclusions, if 'All C_A is B' is true. This class of arguments could be said to be 'truth-preserving to degree 1.00'.

16. In contrast to the above class of probability arguments we could, indeed, have considered the stronger class

(3/4 C_A is B) & Q		(3/4 C_A is B) & Q
So (probably to degree	...	So (probably to degree
.75) a_1 is B		.75) a_n is B

thus making use of the quantitative concept of probability. But the arguments we have actually been considering are sufficient for our present purposes.

17. Now, we might be tempted to generalize this line of thought, by construing the soundness of all probabilistic arguments

g & Q
So (probably) p

in terms of the idea that each such argument, if sound, belongs to a class of argument types of which it can be shown that the majority have true conclusions if their premisses are true.

18. If we characterize an argument as 'successful' if its premisses and conclusion are both true, then to take this approach is to tie the concept of 'probabilistic soundness' to *success* in the sense that to know that such an argument is *sound* and its premisses are *true*, is to know that it belongs to a class of arguments, a majority of which are, in this sense, successful.

IV

19. Now in the case of the probabilistic arguments we have been considering, there is an independent check on the success of such arguments. Thus, whereas we *could* use another argument of the same general form to establish that a particular argument (say α) is successful, e.g.,

> 3/4 arguments of the form
> (3/4 C_A is B) & Q
> So (probably) a_i is B
> have true conclusions, if the premiss is true.
> α is an argument of this form.
> The premiss of α is true.
> So (probably) α is successful.

But we can also, given the appropriate observations, argue

> The conclusion of α is '(probably) a_1 is B'
> The premiss of α is '3/4 C_A is B'
> This premiss is true (for we have examined C_A)
> a_1 is B (observation), so 'a_1 is B' is true
> So α is successful.

V

20. Before we turn our attention to probability arguments which do *not* appear to be amenable to this treatment, let us note that it can, with considerable plausibility, be extended to arguments of the form (where '*OA*' stands for *A*'s which have been observed)

> (3/4 OA are B) & Q $\quad$ $OA = \{a_1, \ldots, a_{100}\}$
> So (probably) a_{101} is B.

21. For, let *PA* be the population consisting of $OA + a_{101}$. Clearly, if *AO* has, as specified, 100 members, the proportion of *B*'s in *PA* cannot be less than 75/101. Thus, at least 75/101 of the arguments

> At least (75/101 PA is B) & Q
> So (probably) a_i is B $\quad$ $a_i = \{a_1, \ldots, a_{101}\}$

will have true conclusions if the premiss is true. Therefore, derivatively, at least 75/101 of the arguments

> (75/100 OA is B) & Q $\quad$ $a_i = \{a_1, \ldots, a_{101}\}$
> So (probably) a_i is B

will also have true conclusions if the premiss is true. Consequently, the following would be a derivatively sound probabilistic argument,

(75/100 *OA* is *B*) & *Q*
So (probably to degree at least 75/101) a_{101} is *B*.

22. The account we have been giving of certain probability arguments can, therefore, it would seem, be extended to arguments from a finite number of *observed* cases to an unobserved case. Indeed, by the use of a somewhat more complicated apparatus, it can be extended to arguments of the form

(75/100 *OA* are *B*) & *Q*
So (probably) approximately 75/100 C_A is *B*,

where C_A is a finite class of unobserved *A*'s. For, given a suitable relationship in size between the observed and the unobserved classes of *A*'s, the 'probably' will amount to 'probably to a considerable degree' and the degree of approximation of the proportion of *B*'s in C_A to 75/100 will be high.

23. This extension would account for the soundness (within specifiable limits) of the inference from the *B*-composition of a class of *observed A*'s to the composition of a finite class of *unobserved A*'s.

VI

24. On the other hand, the preceding account would throw no light on the soundness of probabilistic arguments from statements about the composition of finite observed samples to *statistical law-like statements*. If we suppose that a statistical law-like statement has the form

The limit frequency of *B*'s in an infinite series of *A*'s is n/N,

it is clear that the preceding account throws no light on the soundness, if it be sound, of

(n/N *OA* is *B*) & *Q*
So (probably) the limit frequency of *B* in A_∞ is n/N.

And even if we reject this account of the form of a statistical law-like statement, the 'open-endedness' of law-like statements rules out the possibility that their form is akin to those we have been considering. To put it generally, our recent discussion throws no light on arguments of the form

$(n/N\ OA$ is $B)\ \&\ Q$
So (probably) $LL_{A,\,B,\,n/N}$,

where '$LL_{A,\,B,\,n/N}$' represents a law-like statement, whatever its proper form, which relates the proportion of B's among A's to the ratio n/N.

25. It does not seem possible to construe the soundness of probabilistic arguments of the latter form as a matter of their being members of classes of arguments of which it can be shown that a majority have true conclusions, if their premisses are true, i.e., are successful, if *their* premisses are true.

26. The possibility arises, therefore, that the account we have been giving of the soundness of arguments of the form

$(3/4\ C_A$ is $B)\ \&\ Q$
So (probably) a_i is B

contains a fundamental error, though enough truth to keep the error from bursting out into the open. To gain perspective, it will be useful to consider an ostensible type of probabilistic argument with respect to which – if it be recognized at all – quite a different account *must* be given. If we take seriously the idea that there is such a thing as the probability of theories, where by 'theory' I mean the kind of theory which, as we say, postulates, entities, and processes which are not independently observable.[1] If we take seriously the idea that there is such a thing as the probability of theories in this sense of 'theory', we commit ourselves to the idea that there are good and sound arguments of the form

$g\ \&\ Q$
So (probably) T.

27. What might the grounds represented by 'g' be? Intuitively, we feel that the probability of a theory is a function of its explaining established laws, suggesting new law-like statements capable of empirical tests, and, perhaps, being the simplest (whatever exactly *that* means) of the set of alternatives which are equally effective in the previous respects.

28. Now it might be said that considerations of this kind have *nothing* to do with 'probability' of a theory, but rather with the *reasonableness of accepting it.* Well, they certainly have a lot to do with the latter, but they would *also* have a lot to do with the former, if to say of a theory that it is *probable* were the same thing as to say that *it is* (all things considered) *reasonable to accept it.*[2]

29. Notice that if an argument of the form

> *g* & *Q*
> So, it is reasonable to accept *T*,

were sound and good (though it might be incomplete in the sense that certain additional premisses were 'understood'), then, since it would be reasonable to accept *T* (with the qualifier 'probably') and hence, for example, to inscribe the principles of *T* (with the qualifier 'probably') on a blackboard, it would then be reasonable to inscribe the *pair* of propositions '*g* & *Q*', '(probably) *T*' on the blackboard, thus

> *g* & *Q*
> (Probably) *T*.

Whether it would be appropriate to add 'so', thus

> *g* & *Q*
> *So* (probably) *T*

hinges on whether the '*g* & *Q*' is to be construed as contained among the premisses of an argument of which the conclusion is '(probably) *T*'. For if the only arguments which establish the reasonableness of inscribing '(probably) *T*' are arguments which have as their conclusion *not* '(probably) *T*', but rather

> So, it is reasonable to accept *T*.

Then the reasonableness of inscribing the sequence

> *g* & *Q*
> (Probably) *T*

would *not* entitle us to write it as

> *g* & *Q*
> *So* (probably) *T*.

30. As a remote parallel let us note that it is reasonable to accept '2 plus 2=4' *and* reasonable to accept 'The moon is round', and hence to inscribe

2 plus 2=4 The moon is round.	but *not*	2 plus 2=4 *So* the moon is round.

31. If this line of thought is correct, then, even though the sequence

g & Q
(Probably) T

is correct and proper, there is no such thing as a probability *argument* of which the conclusion is

So (propably) T.

32. This suggests the possibility that in *no* case is there a probability *argument* of the form

g & Q
So (probably) p

i.e., that the concept of such probability arguments is an illusion; the division of argument into 'deductive' and 'probability' arguments a mistake. Notice that by a probability argument I mean an argument of which the conclusion is

So (probably) p

which *asserts p*, though in a qualified way. I do not mean to say that there are no probability arguments, if by this is meant an argument which has as its conclusion

So it is probable that p.

The latter conclusion does not assert 'p'; it asserts a higher order proposition about 'p' – perhaps the higher order proposition that it is reasonable to assert that $-p$.

VII

33. Notice that correlated with every good deductively valid argument, for example,

> (If *p*, then *q*) & *p*
> So (necessarily) *q*

is a certain kind of second order argument, thus (roughly)

> '(*p*, then *q*) & *p*' implies '*q*'
> '(If *p*, then *q*) & *p*' is true
> So it is (epistemically) reasonable to accept '*q*'.

This second-order argument has a 'deductive' flavor which is worth savoring. Indeed, we notice that it would be deductively valid, if we add the premiss

> It is ε-reasonable to accept propositions which are implied by true propositions.

What about this premiss? It might be thought that it is a tautology; in particular that

> '*q*' is implied by a true proposition

has the same sense as

> It is ε-reasonable to accept '*q*'.

Yet it does not seem so. Implication is essentially truth-preservingness, whereas the latter concerns the reasonableness of doing something, accepting a proposition.

34. Notice that if

> '*q*' is true

meant the same as

> It is ε-reasonable to accept '*q*',

then truth-preservation would amount to preserving 'ε-reasonableness of acceptance', and

> '*q*' is implied by a true proposition

would entail

> It is ε-reasonable to accept '*q*'.

35. But

'*q*' is true

does not mean the same as 'It is ε-reasonable to accept "*q*"', for it can be ε-reasonable for Jones to accept '*q*' even though '*q*' is, in point of fact, false.

36. On the other hand, if we are prepared to say something like

'*q*' is true = it is *ideally* ε-reasonable to accept '*q*'

and to distinguish *ideal* ε-reasonableness from ε-reasonableness *for Jones at time t*, then we could put the previous point by saying that while it cannot be *ideally* epistemically reasonable to accept a false proposition, it can be ε-reasonable (for Jones, at *t*) to accept it.

37. It might be pointed out that while we are able to distinguish between the *concept* of a true proposition and the *concept* of a proposition which it is ε-reasonable (for me, now) to accept, statements of the form

Although '*p*' is true, it is not ε-reasonable (for me, now) to accept '*p*'

are *pragmatically* absurd. Thus, in a certain sense of 'imply',

'"*p*" is true' *implies* 'It is ε-reasonable (for me, now) to accept "*p*"'.

38. From this point of view

'*q*' is implied by a true proposition

implies (though it is not synonymous with)

It is ε-reasonable (for me, now) to accept '*q*'

i.e., is, in this particular case a *ground* for the latter.

39. If, following this line of thought, we distinguish between the '*ground*' of its being ε-reasonable (for me, now) to accept '*q*' and the *meaning* of 'It is ε-reasonable (for me, now) to accept "*q*"' (as we distinguish between the ground of an obligation, e.g.,

I promise to do *A*

and the meaning of 'It is morally reasonable for me to do *A*', then we see that there might be *other* grounds for its being ε-reasonable (for me, now) to accept a proposition, than the fact that it is implied by a true proposition.

40. Indeed we have already considered such another gound in our discussion of the proportional syllogism

(3/4 C_A is B) & Q
So (probably) a_i is B.

A moment ago we paralleled the deductive argument

(If p, then q) & p
So (necessarily) q

with

'(If p, then q) & p' implies 'q'
'(If p, then q) & p' is true.
So it is ε-reasonable (for me, now) to accept 'q'.

Let us therefore construct the higher order counterpart of what we have been construing as a probability argument with the premiss '(3/4 A is B) & Q'. It would be something like

'a_i is B' stands in R_L to '(3/4 C_A is B) & Q'
'(3/4 C_A is B) & Q' is true
So it is ε-reasonable (for me, now) to accept 'a_i is B'.

41. In our previous discussion we construed R_L as the relation of

> Being the conclusion of an argument belonging to a family of arguments 3/4 of which have true conclusions if their premisses are true.

But in the light of our discussion of the probability of theories, it occurs to us that this might be a mistake, not because something *like* this relation does not obtain, but because (a) a more basic relation obtains, and (b) this more basic relation insures that 3/4 of the *pairs of statements* – not necessarily *arguments*, though they were previously construed as arguments –

> (3/4 C_A is B) & Q
> (Probably) a_1 is B
> ⋮
> (3/4 C_A is B) & Q
> (Probably) a_n is B

have true second members, if the first member is true.

42. The relation in question between

> 'a_i is B' and '(3/4 C_A is B) & Q'

consists in the fact that it is a logical truth that

> If '3/4 C_A is B' is true, then 3/4 {'a_1 is B', 'a_2 is B', ..., 'a_n is B'} are true.

43. This suggests that the rational of accepting the statement-pair

> (3/4 C_A is B) & Q
> (Probably) a_i is B

rests on a rationale for accepting 'a_i is B' *which is not a matter of '(probably) a_i is B' being the conclusion of an argument*. The suggestion works out neatly as follows

> {It is ε-reasonable (for me, now) to accept all members of a set of propositions 3/4 of which are true, if Q}
> 3/4 {'a_1 is B', 'a_2 is B', ..., 'a_n is B'} are true
> Q
> So {it is ε-reasonable (for me, now) to accept all members of 'a_1 is B', 'a_2 is B', ..., 'a_n is B'.}

Both the reasonableness expressed by the major premiss and the reasonableness of accepting the deductive implications of true premisses would seem to concern the reasonableness of possessing a stock of propositions which are either true or within which a rationally controlled proportion are false.

44. If this interpretation is correct, there is no such thing as an argument

> (3/4 C_A is B)
> So (probably) a_i is B

i.e., no such thing as a non-deductive probability argument. The argument by virtue of which it is reasonable to accept '(Probably) a_i is B' has as its conclusion not '(Probably) a_i is B', but rather

It is ε-reasonable (for me, now) to accept 'a_i is B'

and the argument of which *this* is the conclusion is a *deductively* valid argument.[3]

VIII

45. Suppose all this is correct, what is the larger framework into which it fits? What are the broader implications of the concept of epistemic reasonableness (for s, at t)? In the first place, the epistemic reasonableness in question is, as we saw, that of accepting a proposition. If, for the moment, we construe 'accepting a proposition' as 'bringing it about that one believes the proposition', we tie in with the concept of the reasonableness of an action.[4] To say that an action is (all things considered) reasonable is (in first approximation) to say that one has, all things considered, a good argument for doing it. This argument will be of the kind traditionally called practical. Such arguments have a variety of forms of which the following is an illustration

(All things considered) I shall bring about E.
Bringing about E implies doing A at t, if in circumstances C at t.
I am in C now.
So I shall do A.

The phrase 'all things considered' is essential, *for it is the root of the principle of total evidence.*

46. The above argument contains a 'practical' premiss ('I shall bring about E'), two matter-of-factual premisses of quite different kinds ('bringing about E implies doing A_i at t, if in C_j at t' and 'I am in C now, at t' and a practical conclusion ('I shall do A'). It is clear that the skeleton of the argument is deductive, as can be seen by considering the argument

Jones brought about E.
Bringing about E implies doing A_i at t, if in C_j at t.
Jones was in C_j at t.
So Jones did A_i.

But just how this deductive skeleton is involved in the specific nature of practical reasoning is a complicated topic which I have explored on other occasions.[5]

47. We saw that a *deductive* argument is good, if it is (a) valid, (b) its premisses (and hence its conclusion) are true. The argument about Jones would be a *good* one if its premisses were true. What about the original *practical* argument?

48. Insofar as its factual premisses are true, it can be said to be 'factually' good. But its practical premiss, 'I shall bring about E', is, on the desired interpretation, neither true nor false. It has a force akin to 'Would that I brought about E' or 'Let me bring about E'. The same is true of the conclusion.

49. The argument could be said to be *good without qualification* if 'statements of intention', though neither true nor false, had a character analogous to 'truth' – 'practical objectivity' we might call it, or even 'practical truth'. Thus if

(All things considered) I shall promote E

had 'practical truth', then the argument

All things considered I shall promote E.
Promoting E implies doing A_i at t, if in C_j at t.
I am in C_j now.
So I shall do A_i.

would be both *practically* 'good' as well as *factually* 'good'.

50. If a practical argument is *practically* good, its conclusion will be said to be *categorically reasonable*. If it is merely *factually* good, its conclusion will be said to be *hypothetically reasonable* or reasonable relatively to the purpose expressed by the major premiss.

51. If the argument is merely factually good, thus:

(All things considered) I shall poison my aunt today.
The only available poison is this prussic acid.
......
So I shall give her this prussic acid.

establishes what has been called a 'hypothetical imperative' i.e.,

If (all things considered) I want to poison my aunt today, then I ought to give her this prussic acid.

But a practical argument which is merely *factually* good can never generate an unhypothetical or categorical 'ought'.

52. Otherwise put, a practical argument which is merely factually good can generate:

> If (all things considered) I want to bring about E, it is reasonable (for me, now) to do A,

but it can never lead to

> It is categorically reasonable (for me, now) to do A

even if it is true that all things considered I want to bring about E, i.e., by the use of *modus ponens*.

53. Now I have claimed that the argument which establishes the ε-reasonableness of accepting (i.e., which establishes the probability of) law-like statements, theories, singular statements, etc., are deductive arguments which have as their conclusion

> (All things considered) it is ε-reasonable (for me, now) to accept 'p'; that is (All things considered) it is probable that p.

54. The phrase 'ε-reasonable' carries with it the implication that the end in terms of which the accepting of 'p' is justified by such arguments is an epistemic one, possible, let us say, the possession by some person or persons of a maximum of truth.

55. Given that the following of certain policies can be shown to be a necessary condition of achieving this end, there would be *available* factually good practical arguments of the form:

> (All things considered) I shall promote the possession by such-and-such a person or persons of a maximum of truth (E).
> Promoting E implies doing A_i at t, if in C_j at t.
> I am in C_j.
> So I shall do A_i (e.g., accept a certain theory, etc.).

56. If I actually reason this way, my conclusion can be said to be reasonable relatively to the epistemic end in question.

57. But can we remain satisfied with the idea that the reasonableness of accepting law-like statements, theories, singular statements, etc., is simply a function of an end one *happens* to have? Thus, suppose I simply *like* promoting the truth. I could say:

> If I want to promote the truth, I ought to accept '*p*'.

But the mere fact that I *want* to promote the truth – enjoy the idea of promoting it – could never generate the categorical

> I ought to accept '*p*'.

58. There are two considerations which lead me to suppose that *wanting* to promote truth cannot be the end of the story. In the first place, probability statements are *intersubjective*. If Jones says to Smith, 'It is probable that *p*' and Smith agrees, they are agreeing about the same thing. This suggests that

> It is probable that *p*

has something like the sense of

> It is ε-reasonable (for *us*, now) to accept '*p*'.

In the second place whenever a person acts on a probability, he regards the kind of action he decides to do as *reasonable* because he thinks that a substantial series of such actions in the kind of circumstance in which the contingency may or may not occur would probably maximize relevant values or utilities. But, as Peirce pointed out, we regard such actions as reasonable even when we know that we *as individuals* will not be in that kind of circumstance often enough to make this consideration relevant and, to put an extreme case, we regard such actions as reasonable *even when we know that we are about to die*. Peirce concludes, correctly, I think, that in thinking and acting in terms of probability, we are, in a certain sense, identifying ourselves with a continuing community.

59. Both of these considerations suggest that the prime mover of the practical reasoning involving in probabilistic thinking, indeed all logically oriented thinking, 'deductive' as well as 'inductive', 'practical' as well as 'theoretical', is not an idiosyncratic *wish* to promote truth, but the intention *as a member of a community* to promote the total welfare of that community. This intention is implicit, and, when it becomes explicit, can be overcome by impulse and self-interest. But it is because truth is a necessary condition of securing the common good that the search for it presents itself to us, on reflection, as categorically reasonable – in the truest sense a moral obligation.

University of Pittsburgh

REFERENCES

[1] I say 'not independently observable' because there is a dependent sense in which, e.g., the motion of an electron is observable – cf. gas and bubble chambers – but observable in a sense which involves a tacit appeal to the theory.

[2] This use of 'accept' must be distinguished from the ordinary sense in which one 'accepts', for example, the statements (testimony) of others. I use 'accept', in the first instance, as roughly equivalent to 'come to believe'. Believing and coming to believe are things that one does only in that broad sense in which anything expressed by a verb not in the passive voice is a 'doing'.

[3] For a more detailed account of probability statements and arguments which explores the inter-relationships between the various modes of probability see my 'Induction as Vindication', *Philosophy of Science* **31** (1964), pp. 197–231.

[4] A finer grained analysis of the context 'It is reasonable to accept that-*p*' would require careful use of the distinction between *rules of action* and *rules of criticism* developed, among other places, in 'Language as Thought and as Communication', *Philosophy and Phenomenological Research* **29** (1969), pp. 528–535.

[5] Most recently in Chapter VII of *Science and Metaphysics*, Routledge and Kegan-Paul, London, 1968.

RICHARD C. JEFFREY

STATISTICAL EXPLANATION VS. STATISTICAL INFERENCE

Hempel is not the first philosopher to have held that causal explanations are deductive inferences of a special sort: in the *Posterior Analytics*[1] Aristotle distinguishes a special sort of deductive inference – the demonstrative syllogism – in these terms:

By demonstration I mean a syllogism productive of scientific knowledge, a syllogism, that is, the grasp of which is *eo ipso* such knowledge.

He then lays down defining conditions for this special sort of inference:

... the premisses of demonstrated knowledge must be true, primary, immediate, better known than and prior to the conclusion, which is further related to them as effect to cause.

And he remarks,

Syllogism there may indeed be without these conditions, but such syllogism, not being productive of scientific knowledge, will not be demonstration.

Now we can fault this account on various grounds, but so can we fault contemporary accounts. We must give the old man credit; as he says at the end of the *Organon* (at the end of *De Sophisticis Elenchis*), his was the first book on logic; and he concludes,

... there must remain for all of you, or for our students, the task of extending us your pardon for the shortcomings of the inquiry, and for the discoveries thereof your warm thanks.

The affinities between the Hempelian and Aristotelian accounts of explanation may be obscured by differences in terminology. Thus, Aristotle speaks of syllogism, Hempel of deductive inference; and Aristotle speaks of knowledge, Hempel of explanation. But remember that 'syllogism' was Aristotle's general term for deductive inference – and do not tax him with what we now know to be his overly narrow view of the forms that such inferences can take. As to knowledge vs. explanation, Aristotle says

We suppose ourselves to have unqualified scientific knowledge of a thing, as opposed

N. Rescher et al. (eds.), Essays in Honor of Carl G. Hempel.

to knowing it in the accidental way in which the sophist knows, when we think that we know the cause on which the fact depends ....

It is precisely this sort of understanding that is conveyed by causal explanation – 'the grasp of which is *eo ipso* such knowledge'. Then I take it that Aristotle's *demonstrated knowledge* – knowledge, as he says later, not simply of the fact but of the *reasoned fact* – is the psychological correlate of Hempel's causal explanation as it is of Aristotle's demonstrative syllogism.

Of course, I have been overemphasizing the similarities between the Aristotelian and Hempelian accounts of scientific understanding; and of course, my point is not to advocate a back-to-Aristotle movement. Still less do I seek to belittle Hempel's work on explanation; the point is rather to appreciate that work as continuous with the philosophical enterprise that started in Athens 2400 years ago.

My real concern here is not with the history of nomological-deductive accounts of explanation – interesting as that may be – nor is it with the nomological-deductive view of causal explanation itself – satisfactory as *that* may be. My concern is with statistical explanation, and I have begun by recalling a general feature of nomological-deductive explanation because I want to show that statistical explanations often lack that feature. The particular feature is that of *being an inference*. On Aristotle's account as on Hempel's, causal explanations are deductive inferences; and it seems plausible to say that for their part, statistical explanations are statistical inferences – or, if you prefer, inductive or probabilistic inferences. But I shall argue that only in certain cases is it plausible to view statistical explanations as statistical inferences.[2]

The really beautiful cases of statistical explanation are the statistical mechanical ones in which the explained occurrence can be shown to have probability so close to 1 as to 'make no odds' in any gamble or other decision problem. I am thinking here of such cases as the explanation of why a flat tire does not spontaneously inflate: it is *possible* that the random movements of the surrounding air molecules might, for a few seconds, be such as to constitute a jet of air through the puncture powerful enough to pump the tire up – possible, but so improbable as to make it a *practical certainty* that no such jet will be forthcoming. (A proposition is a practical certainty if its probability is so high as to allow us to reason, in *any* decision problem, as if its probability were 1.) It is in these beautiful, extreme cases that the view of statistical explanations as statistical inferences is no less plausible than the view of causal explanations as deductive inferences.[3]

But there is a certain obliqueness about these inferential explanations, whether the strength of the inference be deductive certainty or practical certainty. I mean, one explains in order to impart knowledge of *how* or *why* the explained phenomenon takes place, but the explanation itself (in these cases) takes the form of a proof that the phenomenon *does* take place. The explanation is just the sort of thing one might produce in order to prove *that* the explained phenomenon takes place; this is the famous parallelism between explanation and prediction which I think breaks down for statistical explanations that impart less than practical certainty to the phenomenon explained.

I say that causal explanations (and the 'beautiful' kind of statistical explanations) are oblique: they explain *how* or *why* by demonstrating *that* in a special way. (Compare Aristotle's conception of knowledge of the *reasoned* fact.) Of course, not every demonstration *that* serves to explain *why*, as Louis MacNeice's rhyme testifies:

> The glass is falling hour by hour, the glass will fall forever.
> But if you break the bloody glass you won't hold up the weather.[4]

The falling of the glass does not cause the weather to turn bad; it is rather an effect of something that the bad weather is also an effect of. Suppose it were true that *whenever* the glass falls, the weather turns bad. Then the inference

The glass is falling.
Whenever the glass falls the weather turns bad.

∴ The weather will turn bad.

would support a prediction *that* the weather will turn bad without explaining *why* the weather turns bad. On the other hand, the inference

The glass is falling.
Whenever the glass is falling the atmospheric pressure is falling.
Whenever the atmospheric pressure is falling the weather turns bad.

∴ The weather will turn bad.

would serve both to predict and explain the weather: it gives us knowledge not merely of the fact, but of the reasoned fact, for it proves that the fact is a fact by citing causes and not mere symptoms. In general, where the inference meets certain conditions, one of which is that a causal law appear among the premisses, deductively grounded prediction will double as explanation. In these cases, knowledge *that* is imparted in such a way as to provide knowledge *why* as well. I mention this in order to suggest that this way of telling why obliquely, by demonstrating that, may not be the only way or the best way of telling why: the fact that we can sometimes explain by inferring is not a very strong reason to suppose that we can *only* explain by inferring. Perhaps, then, we should have another look at causal explanations, to see whether they *can* always be cast in the Aristotelian-Hempelian mould; but that is for another occasion. Here I want to have a look at statistical explanations, where I think it can pretty readily be seen that many of them simply cannot be represented as statistical inferences which show *why* obliquely, by demonstrating *that.*

Consider the following set of coin-tossing examples. A fair coin is tossed n times ($n=1, 2, 3, \ldots$), and at least one head turns up. Here the Hempelian statistical inference would have as its premiss a statistical law indicating that the probability of a head on any toss is $\frac{1}{2}$ and that distinct tosses are statistically independent, and would have as its conclusion the statement $H_1 \vee H_2 \vee \cdots \vee H_n$ that at least one of the first n tosses yields a head. The premiss – the statistical law – might be specified in various equivalent ways, e.g., by saying that the probability of the conjunction of any k distinct H's is $\frac{1}{2}^k$. From this specification one can deduce, via the laws of the elementary probability calculus, the statistical probability of any truth functional compound of the H's, e.g., the statistical probability of the conclusion of the statistical inference which Hempel represents as follows:

$$\frac{\text{If } i_1, \ldots, i_k \text{ are all distinct, then } p(H_{i_1} \,\&\cdots\&\, H_{i_k}) = \frac{1}{2}^k}{H_1 \vee H_2 \vee \cdots \vee H_n} \; [1 - 2^{-n}]$$

The bracketed number at the right represents the inductive probability – the degree of confirmation – of the conclusion, conditionally upon the premiss. If we represent the premiss by 'P' and the conclusion by 'C' then the foregoing array is to be viewed as a typographical variant of the statement

$$c(C, P) = 1 - 2^{-n}$$

which gives $1-2^{-n}$ as the inductive probability or degree of confirmation of the conclusion on the premiss.[5]

Hempel would hold that with $n=1$ the foregoing inductive inference is too weak to serve as an explanation of the fact that there was at least one head on the first n tosses: with $n=1$ the bracketed number – what Hempel calls *the strength of the explanation* – is only $\frac{1}{2}$. At the other extreme – for such large values of n as 100 – even I will admit that the inference will serve as an explanation, for a gap of $\frac{1}{2}^{100}$ between the probability of the conclusion and 1 is so small as to make no odds in any deliberation. But Hempel might hold that even with $n=10$ we have an explanation: the bracketed 'strength' of the inference is then 1023/1024 or a bit over .999.

Now I agree that with $n=10$ the strength of the inference is great enough to justify giving long odds, on the order of 1000:1, on there being at least one head. The number 1023/1024 is a good measure of the proper strength of our expectation of the fact; but I deny that it is a good measure of the quality or strength of the explanation which the inference gives us. Indeed, I think it misleading to think of the statistical inference as being an explanation at all. The explanation how or why there was at least one head, I should think, would be given by specifying the statistical probability function p that governs the process, e.g., as in the premiss of the inference, above. Perhaps it *is* part of the explanation to point out (as in the brackets above) that the probability of the explained phenomenon is then $1-(\frac{1}{2}^n)$, *but the strength of the explanation would be no greater with* $n=10$ *than with* $n=1$! To say, in the case $n=1$, that the statistical probability was $\frac{1}{2}$, is as much of an explanation as one can give. Indeed, this is a causal explanation in a sense in which a proof that there is at least one head can never be, for to say that the statistical probability is $\frac{1}{2}$ is to say *directly* something about cause which is said only obliquely, if at all, in a proof that there is at least one head. What is being said about cause, when we say that the statistical probability is $1-(\frac{1}{2}^n)$ that there will be at least one head, is that this effect is 'caused' by a chance or random process: we are saying directly that the usual sort of causality is absent.

To see this point it is helpful to consider what we should say if the improbable happened, say in the case $n=2$, and there were two tails. The strength of the 'explanation' of $-H_1$ & $-H_2$ would then be $\frac{1}{4}$ on Hempel's account, but we still have as complete an understanding of the why and the how of the outcome as we would have had if a head had turned up and the outcome had been the more probable one.

Moral: The strength of a statistical explanation (except in Hempel's technical sense) is not given by the degree of confirmation that the premisses bestow on the conclusion in the corresponding Hempelian inductive inference. Similarly for $n=10$: it is possible, although highly unlikely, that there will be ten tails, and if this happens we shall know all there is to know about the why of it and the how, when we know that the process which yielded the ten tails is a random one and when we know the probabilistic *law* governing the process. The knowledge that the process was random answers the question, 'Why?' – the answer is, 'By chance'. Knowledge of the probabilistic law governing the process answers the question 'How' – the answer is, 'Improbably, as a product of such-and-such a stochastic process'. Note that knowledge of the probabilistic law of the process makes it at most a matter of calculation to find that the statistical probability of the phenomenon – ten tails – is 1/1024. To know the probabilistic law is to know, among other things, that the actual outcome was hardly to have been expected.

Here knowledge *why* splits clearly away from knowledge *that.* Indeed in the statistical case I find it strained to speak of knowledge *why* the outcome is such-and-such. I could rather speak of *understanding the process* which had the outcome, for the explanation is basically the same no matter what the outcome: it consists of a statement that the process was a stochastic one, following such-and-such a law. (One may gloss this statement by pointing out that the actual outcome had such-and-such a probability, given the law of the process; but this gloss is not the heart of the explanation.)

Put the matter in this way: Aristotle's *knowledge of the reasoned fact* just will not do in the statistical case, for where a statistical explanation is appropriate, there's *no* reason for the fact: it came about by chance. Nor is the situation changed when (as usual) the probable happens. Because such cases are usual, we can usually give a statistical *inference* of strength $\frac{1}{2}$ or more then we can give a statistical explanation; and this, I take it, is why it is easy to mistake the inference for an explanation.

That the inference is *not* an explanation is shown, I think, by the fact that even when the improbable chances to happen, we give the same sort of account: the happening was the product of a *stochastic* process following such-and-such a probabilistic law. And we gloss this by pointing out that in this case the unexpected happened. My point is that it is no less a gloss, and no more essentially a part of the explanation, when we point out in the more usual cases that the *expected* happened.

Let us conclude by examining some cases in which explanation can be both causal and statistical. *Why was my first child a boy*? There are two sorts of answers: the statistical one ('There is no damned reason – it was pure chance'), and the causal ('Because the germ cell I contributed to the zygote which developed into the child was of the *Y* genotype'). Both answers are right; they supplement each other. The full-blown causal explanation might look like this:

> The sperm which united with an ovum to form the zygote out of which my first child developed had the *Y* genotype.
> Whenever the sperm which unites with an ovum to form the zygote out of which a child develops has the *Y* genotype, the child is a boy.
>
> ---
> My first child was a boy.

Here the first premiss could not have been known to be true before the birth, and so the inference could not have been used to support a prediction of the sex of the child; but it is a perfectly satisfactory causal explanation for all that. The second premiss of the explanatory inference is a causal law. Its converse is also true, and can be used in conjunction with the conclusion of the explanatory inference to deduce the first premiss of that inference. Indeed, that is how I know the genotype of the successful germ cell! But the deduction which tells me that is no part of the explanatory inference; it is rather part of the business of verifying that the explanatory inference really does meet the conditions for being a causal explanation. ('The premisses of demonstrated knowledge must be true' among other things.) The causal and statistical explanations can be made to match by referring the statistics not to the sex of the child, but to the genotype of the relevant sperm cell: the usual process whereby spermatozoa unite with ova is a stochastic one in which the statistical probability is 1/2 that the winner will have the *Y* genotype.

This is a case where before the outcome is known, we know that one of two causal explanations of the outcome will be correct – and we shall know *which* explanation is correct when we know the outcome. Before the event we had a lottery with causal explanations as prizes.

The same sort of thing can happen after the event, as in this made-up example: I draw at random from a box containing a two-headed penny and a normal one, toss, observe that a head turned up, and return the penny to the box without examining it further. Here the lottery is between a statistical and a causal explanation; the outcome of the lottery is not settled by the outcome of the toss, as it would have been if a tail had turned up; and we end in a position where as far as we know, the question 'Why' may or may not have an answer!

Finally, consider the question of why my car would not start this morning. Here the question surely has an answer: there *is* a causal explanation although I do not know what it is. Suppose that it is only one morning in about twenty that my car will not start, and suppose that when it *does not* start, then nine times out of ten it is because of the defective frammas which I never had fixed. Here the statistical component generates an epistemic lottery in which the prizes are causes. There is an answer to the question 'Why?' and the statistics give us a strong clue: it is very likely to be the frammas. Indeed, the probability that it is the frammas is .9, but there is no temptation here to speak of a statistical explanation of strength .9. Rather, the statistical component of the explanation goes like this: non-starting is the product of a stochastic process according to which the frammas goes bad with probability (1/20) (9/10) = .045 and something else goes bad with probability (1/20) (1/10) = .005, and in either case the car definitely is prevented from starting. Given this stochastic law, the probability of non-starting is only .05, and this would be the strength of the Hempelian inferential explanation of the non-starting. But the description of the stochastic process is better than the number .05 would indicate, as a statistical explanation: as far as it goes, it is perfect, and it goes as far as statistics will take us. What is needed to round it out is an item of causal knowledge: was it the frammas?

CONCLUSION

Sometimes statistics enter our understanding of phenomena *via* causal lotteries. In such cases the full explanation would have both a statistical and causal component; and it is doubtful whether Hempel would want to apply his statistical inferential model in such cases. Among the cases where he would apply his model, some are of the 'beautiful' variety where the strength of the inference is so great as to allow us to speak of practical certainty in this sense: the probability of the phenomenon to be explained is so high, given the stochastic law governing the process that produces it, as to make no odds in any gamble or deliberation. These are the cases where the inferential model of explanation seems unexceptionable. But where the strength of the inference is more modest, I think it simply wrong to view the inference as an explanation, and to identify the strength of the inference with the strength of the explanation. To explain the phenomenon that there was at least one head in two tosses of a coin, I would point out that the process is stochastic with probability $\frac{1}{2}$ of head on each toss, and with different tosses independent of one another. I would give the same explanation if matters turned out differently: if, improbably, there had been no head on either toss. The difference between the two cases would lie entirely with the gloss: in the first case one would point out that the probable happened, while in the second one would point out that the improbable happened. But the strength of the explanation would be the same in either case. Finally I point out that since the probable happens more often than not, we are usually able to provide a Hempelian inference when we are able to give a statistical explanation; and I suggest that this is what gives the view that statistical explanations are statistical inferences its specious plausibility.

University of Pennsylvania

REFERENCES

[1] Book 1, chapter 2. All citations from this work are from the Oxford translation.
[2] I leave to one side the deductive-statistical explanations discussed in Part 3.2 of *Aspects of Scientific Explanation*, The Free Press, New York, 1965.
[3] The notion of practical certainty operative in this account of the 'beautiful' cases of statistical explanation is problematical but, I think, defensible. Example: There would be serious trouble with that notion if, given a probability greater than 0 (no matter how slightly greater), one could name a prize so great that the prospect of getting that prize with that probability is not negligible. But I take it that such problems as the St. Petersburg paradox already force us to realize that if Bayesian decision-theory

is to work, there must be a finite upper bound on the utilities of the things the agent can envisage as prizes.

[4] The lines from Louis MacNeice are the last two of his 'Bagpipe Music'.

[5] I would prefer to avoid this second kind of probability and speak simply of the statistical probability of the conclusion; indeed, $p(C)$ is $1-2^{-n}$, i.e., the probability measure defined in the premiss assigns to the conclusion precisely the value which the inductive probability measure c assigns to the conclusion conditionally on the premiss. But let us stay with Hempel's way of talking.

ROBERT NOZICK

NEWCOMB'S PROBLEM AND TWO PRINCIPLES OF CHOICE*

> Both it and its opposite must involve no mere artificial illusion such as at once vanishes upon detection, but a natural and unavoidable illusion, which even after it has ceased to beguile still continues to delude though not to deceive us, and which though thus capable of being rendered harmless can never be eradicated.
>
> IMMANUEL KANT, *Critique of Pure Reason*, A422, B450

I

Suppose a being in whose power to predict your choices you have enormous confidence. (One might tell a science-fiction story about a being from another planet, with an advanced technology and science, who you know to be friendly, etc.) You know that this being has often correctly predicted your choices in the past (and has never, so far as you know, made an incorrect prediction about your choices), and furthermore you know that this being has often correctly predicted the choices of other people, many of whom are similar to you, in the particular situation to be described below. One might tell a longer story, but all this leads you to believe that almost certainly this being's prediction about your choice in the situation to be discussed will be correct.

There are two boxes, (B1) and (B2). (B1) contains \$1000. (B2) contains either \$1000000 (\$*M*), or nothing. What the content of (B2) depends upon will be described in a moment.

$$(\text{B1})\ \{\$1000\} \qquad (\text{B2}) \begin{Bmatrix} \$M \\ \text{or} \\ \$0 \end{Bmatrix}$$

You have a choice between two actions:

(1) taking what is in both boxes
(2) taking only what is in the second box.

N. Rescher et al. (eds.), Essays in Honor of Carl G. Hempel.

Furthermore, and you know this, the being knows that you know this, and so on:

(I) If the being predicts you will take what is in both boxes, he does not put the $M in the second box.

(II) If the being predicts you will take only what is in the second box, he does put the $M in the second box.[1]

The situation is as follows. First the being makes its prediction. Then it puts the $M in the second box, or does not, depending upon what it has predicted. Then you make your choice. What do you do?

There are two plausible looking and highly intuitive arguments which require different decisions. The problem is to explain why one of them is not legitimately applied to this choice situation. You might reason as follows:

First Argument: If I take what is in both boxes, the being, almost certainly, will have predicted this and will not have put the $M in the second box, and so I will, almost certainly, get only $1000. If I take only what is in the second box, the being, almost certainly, will have predicted this and will have put the $M in the second box, and so I will, almost certainly, get $M. Thus, if I take what is in both boxes, I, almost certainly, will get $1000. If I take only what is in the second box, I, almost certainly, will get $M. Therefore I should take only what is in the second box.

Second Argument: The being has already made his prediction, and has already either put the $M in the second box, or has not. The $M is either already sitting in the second box, or it is not, and which situation obtains is already fixed and determined. If the being has already put the $M in the second box, and I take what is in both boxes I get $M + $1000, whereas if I take only what is in the second box, I get only $M. If the being has not put the $M in the second box, and I take what is in both boxes I get $1000, whereas if I take only what is in the second box, I get no money. Therefore, whether the money is there or not, and which it is already fixed and determined, I get $1000 more by taking what is in both boxes rather than taking only what is in the second box. So I should take what is in both boxes.

Let me say a bit more to emphasize the pull of each of these arguments:

The First: You know that many persons like yourself, philosophy teachers and students, etc., have gone through this experiment. All those who took only what was in the second box, included those who knew of the second argument but did not follow it, ended up with $\$M$. And you know that all the shrewdies, all those who followed the second argument and took what was in both boxes, ended up with only \$1000. You have no reason to believe that you are any different, *vis-à-vis* predictability, than they are. Furthermore, since you know that I have all of the preceding information, you know that I would bet, giving high odds, and be rational in doing so, that if you were to take both boxes you would get only \$1000. And if you were to irrevocably take both boxes, and there were some delay in the results being announced, would not it be rational for you to then bet with some third party, giving high odds, that you will get only \$1000 from the previous transaction? Whereas if you were to take only what is in the second box, would not it be rational for you to make a side bet with some third party that you will get $\$M$ from the previous transaction? Knowing all this (though no one is actually available to bet with) do you really want to take what is in both boxes, acting against what you would rationally want to bet on?

The Second: The being has already made his prediction, placed the $\$M$ in the second box or not, and then left. This happened one week ago; this happened one year ago. Box (B1) is transparent. You can see the \$1000 sitting there. The $\$M$ is already either in the box (B2) or not (though you cannot see which). Are you going to take only what is in (B2)? To emphasize further, from your side, you cannot see through (B2), but from the other side it is transparent. I have been sitting on the other side of (B2), looking in and seeing what is there. Either I have already been looking at the $\$M$ for a week or I have already been looking at an empty box for a week. If the money is already there, it will stay there whatever you choose. It is not going to disappear. If it is not already there, if I am looking at an empty box, it is not going to suddenly appear if you choose only what is in the second box. Are you going to take only what is in the second box, passing up the additional \$1000 which you can plainly see? Furthermore, I have been sitting there looking at the boxes, hoping that you will perform a particular action. Internally, I am giving you advice. And, of course, you already know which advice I am silently giving to you. In either case (whether or not I see the $\$M$ in the second box) I am hoping that you will take what is in both boxes. You know that the person sitting and watching it all hopes that you will take the contents of both boxes. Are you going to take only what is in the second box, passing up the additional $1000 which you can plainly see, and ignoring my internally given hope that you take both? Of course, my presence makes no difference. You are sitting there alone, but you know that if some friend having your interests at heart *were* observing from the other side, looking into both boxes, he *would* be hoping that you would take both. So will you take only what is in the second box, passing up the additional $1000 which you can plainly see?

I should add that I have put this problem to a large number of people, both friends and students in class. To almost everyone it is perfectly clear and obvious what should be done. The difficulty is that these people seem to divide almost evenly on the problem, with large numbers thinking that the opposing half is just being silly.[2]

Given two such compelling opposing arguments, it will not do to rest content with one's belief that one knows what to do. Nor will it do to just repeat one of the arguments, loudly and slowly. One must also disarm the opposing argument; explain away its force while showing it due respect.

Now for an unusual suggestion. It might be a good idea for the reader to stop reading this paper at the end of this section (but do, please, return and finish it), mull over the problem for a while (several hours, days) and then return. It is not that I claim to solve the problem, and do not want you to miss the joy of puzzling over an unsolved problem. It is that I want you to understand my thrashing about.

II

My strategy in attacking this problem is ostrich-like; that is, I shall begin by ignoring it completely (except in occasional notes) and proceed to discuss contemporary decision theory. Though the problem is not, at first, explicitly discussed, the course my discussion takes is influenced by my knowledge of the problem. Later in the paper, I shall remove my head from the sand, and face our problem directly, hopefully having advanced towards a solution, or at least having sharpened and isolated the problem.

Writers on decision theory state two principles to govern choices among alternative actions.

> *Expected Utility Principle:* Among those actions available to a person, he should perform an action with maximal expected utility.

The expected utility of an action yielding the exclusive outcomes $O_1, \ldots, O_n$ with probabilities $p_1, \ldots, p_n$ respectively,

$$\left(\sum_{i=1}^{n} p_i = 1\right) \text{ is } p_1 \times u\,(0_1) + p_2 \times u\,(0_2) + \cdots + p_n \times u\,(0_n),$$

$$\text{i.e., } \sum_{i=1}^{n} p_i \times u\,(0_i).$$

> *Dominance Principle:* If there is a partition of states of the world such that relative to it, action *A* weakly dominates action *B*, then *A* should be performed rather than *B*.

Action *A* weakly dominates action *B* for person *P* iff, for each state of the world, *P* either prefers the consequence of *A* to the consequence of *B*, or is indifferent between the two consequences, and for some state of the world, *P* prefers the consequence of *A* to the consequence of *B*.

There are many interesting questions and problems about the framework used or assumed by these principles and the conditions governing preference, indifference, and probability which suffice to yield the utility measure, and the exact way the principles should be formulated.[3] The problem I want to begin with is raised by the fact that for some situations, one of the principles listed above requires that one choose one action whereas the other principle requires that one choose another action. Which should one follow?

Consider the following situation, where *A* and *B* are actions, S_1 and S_2 are states of the world, and the numerical entries give the utility of the consequences, results, effects, outcomes, upshots, events, states of affairs, etc., that obtain, happen, hold, etc., if the action is done and the state of the world obtains.

	S_1	S_2
A:	10	4
B:	8	3

According to the dominance principle, the person should do A rather than B. (In this situation A strongly dominates B, that is, for each state of nature the person prefers the consequence of A to the consequence of B.) But suppose the person believes it very likely that if he does A, S_2 will obtain, and if he does B, S_1 will obtain. Then he believes it very likely that if he does A he will get 4, and if he does B he will get 8.

	S_1	S_2
A:	10	4
B:	8	3

The expected utility of $A = \text{prob}(S_1/A)\,10 + \text{prob}(S_2/A)\,4$. The expected utility of $B = \text{prob}(S_1/B)\,8 + \text{prob}(S_2/B)\,3$. If, for example,

$$\begin{aligned} \text{prob}(S_1/A) &= .2 \\ \text{prob}(S_2/A) &= .8 \\ \text{prob}(S_1/B) &= .9 \\ \text{prob}(S_2/B) &= .1, \end{aligned}$$

then the expected utility of $A = 5.2$, and the expected utility of $B = 7.5$. Thus the expected utility principle requires the person to do B rather than A.[4]

The dominance principle as presented here speaks of dominance relative to a partition of the states of the world. This relativization is normally not made explicit, which perhaps accounts for the fact that writers did not mention that it may be that relative to one partition of the states of the world, one action A dominates another, whereas relative to another partition of the states of the world, it does not.

It will be helpful to have before us two facts:

First: Suppose a matrix is given, with states $S_1, \ldots, S_n$, in which action A does not dominate action B. If there is some rearrangement of the utility entries in the row for action A which gives a new row which dominates the row for action B, then there are states $T_1, \ldots, T_n$ such that in the matrix with these states, action A dominates action B.

Proof: I shall describe how one can get the appropriate states $T_1, \ldots, T_n$ in one case. It is obvious how this procedure can be used generally. Suppose that $a_1, \ldots, a_n$ and $b_1, \ldots, b_n$ are utility numbers such that, for all i, $a_i \geqslant b_i$, and for some i, $a_i > b_i$. We may suppose that a_i is the entry in the A row for the ith column, that is, for state S_i. We might, for example, have the following matrix:

$$\begin{array}{llllll} & S_1 & S_2 & S_3 & \ldots\ldots\ldots\ldots & S_n \\ A: & a_1 & a_2 & a_3 & \ldots\ldots\ldots\ldots & a_n \\ B: & b_{12} & b_3 & b_{19} & \ldots\ldots\ldots\ldots & b_6 \end{array}$$

Let

$$\begin{aligned} T_1 &= A \,\&\, S_{12} \text{ or } B \,\&\, S_1{}^5 \\ T_2 &= A \,\&\, S_3 \text{ or } B \,\&\, S_2 \\ T_3 &= A \,\&\, S_{19} \text{ or } B \,\&\, S_3 \\ &\vdots \\ T_n &= A \,\&\, S_6 \text{ or } B \,\&\, S_n. \end{aligned}$$

Thus we get the matrix,

$$\begin{array}{llllll} & T_1 & T_2 & T_3 & \ldots\ldots\ldots\ldots & T_n \\ A: & a_{12} & a_3 & a_{19} & \ldots\ldots\ldots\ldots & a_6 \\ B: & b_{12} & b_3 & b_{19} & \ldots\ldots\ldots\ldots & b_6 \end{array}$$

In this matrix, action A dominates action B. Since the kind of procedure followed does not depend on any specific features of the example, the point is made.

Second: Suppose there is a matrix with states $S_1, \ldots, S_n$ such that action A dominates action B. If there is some rearrangement of the utility entries in the B row so that the rearranged row is not dominated by A, then there are states $T_1, \ldots, T_n$ such that if the matrix is set up with these states, B is not dominated by A.

Proof: Suppose that $a_i \geqslant b_i$, for all i; $a_i > b_i$ for some i; and that some B-row value is greater than some A-row value. (Given that there is some arrangement in which A dominates B, this last supposition follows from its being possible to rearrange the B row so that it is not dominated by the A row.) Suppose, without loss of generality that $b_{12} > a_2$. Thus we have the following matrix:

$$\begin{array}{llllll} & S_1 & S_2 & S_3 & \ldots\ldots\ldots\ldots & S_n \\ A: & a_1 & a_2 & a_3 & \ldots\ldots\ldots\ldots & a_n \\ B: & b_1 & b_2 & b_3 & \ldots\ldots\ldots\ldots & b_n \end{array}$$

Let

$$
\begin{aligned}
T_1 &= S_1 \\
T_2 &= A \,\&\, S_2 \text{ or } B \,\&\, S_{12} \\
T_3 &= S_3 \\
&\vdots \\
T_{11} &= S_{11} \\
T_{12} &= A \,\&\, S_{12} \text{ or } B \,\&\, S_2 \\
T_{13} &= S_{13} \\
&\vdots \\
T_n &= S_n.
\end{aligned}
$$

Thus we get the following matrix:

	T_1	T_2	T_3	....	T_{12}	....	T_n
A:	a_1	a_2	a_3	....	a_{12}	....	a_n
B:	b_1	b_{12}	b_3	....	b_2	....	b_n

Since $b_{12} > a_2$, A does not dominate B.

It may seem that the states $T_1, \ldots, T_n$ defined in terms of the actions A and B, and the states $S_1, \ldots, S_n$ are contrived states, which some general condition could exclude. It should be noted that – since the states $S_1, \ldots, S_n$ can be defined in terms of the actions A and B and the states $T_1, \ldots, T_n$ (I will give some examples below) – attempts to show that $T_1, \ldots, T_n$ are contrived will face many of the problems encountered in ruling out Goodman-style predicates. Furthermore, as we shall see soon, there are cases where the S states and the T states which are inter-definable in this way, both seem perfectly natural and uncontrived.

The fact that whether one action dominates another or not may depend upon which particular partition of the states of the world is used would cause no difficulty if we were willing to apply the dominance principle to *any* partition of the states of the world. Since we are not, this raises the question of when the dominance principle is to be used. Let us look at some examples.

Suppose that I am about to bet on the outcome of a horserace in which only two horses, H_1 and H_2, are running. Let:

S_1 = Horse H_1 wins the race.
S_2 = Horse H_2 wins the race.
A_1 = I bet on horse H_1.
A_2 = I bet on horse H_2.

Suppose that I will definitely bet on one of the two horses, and can only bet on one of the two horses, and that the following matrix describes the situation. (I might have been offered the opportunity to enter this situation by a friend. Certainly no race track would offer it to me.)

	S_1	S_2
A_1:	I win \$50	I lose \$5
A_2:	I lose \$6	I win \$49

Suppose further that the personal probability for me that H_1 wins is .2, and the personal probability for me that H_2 wins is .8. Thus the expected utility of A_1 is $.2 \times u$ (/I win \$50/) $+ .8 \times u$ (/I lose \$5/). The expected utility of A_2 is $.2 \times u$ (/I lose \$6/) $+ .8 \times u$ (/I win \$49/). Given my utility assignment to these outcomes, the expected utility of A_2 is greater than that of A_1. Hence the expected utility principle would have me do A_2 rather than A_1.

However, we may set the matrix up differently. Let:

S_3 = I am lucky in my bet.
S_4 = I am unlucky in my bet.

(Given that I am betting on only one horse today, we could let S_3 = The only horse I bet on today wins. Similarly for S_4, with 'loses' substituted for 'wins'.) Thus we have the following matrix:

	S_3	S_4
A_1:	I win \$50	I lose \$5
A_2:	I win \$49	I lose \$6

But when set up in this way, A_1 dominates A_2. Therefore the dominance principle would have me do A_1 rather than A_2.[6]

In this example, the states are logically independent of which action I perform; from the fact that I perform A_1 (A_2) one cannot deduce which state obtains, and from the fact that S_1 (S_2, S_3, S_4) obtains one cannot deduce which action I perform. However one pair of states was not probabilistically independent of my doing the actions.[7] Assuming that S_1 and S_2 are each probabilistically independent of both A_1 and A_2, prob (S_3/I do A_1)=.2; prob (S_3/I do A_2)=.8; prob (S_4/I do A_1)=.8; prob (S_4/I do A_2)=.2. Thus neither of the states S_3 or S_4 is probabilistically independent of each of the actions A_1 and A_2.[8]

In this example, it is clear that one does not wish to follow the recommendation of the dominance principle. And the explanation seems to hinge on the fact that the states are not probabilistically independent of the actions. Even though one can set up the situation so that one action dominates another, I believe that if I do A_1, the consequence will probably be the italicized consequence in its row, and I believe that if I do A_2, the consequence will probably be the italicized consequence in A_2's row. And given my assignment of utilities in this case, and the probabilities I assign (the conditional probabilities of the states given the actions) it is clear why I prefer to do A_2, despite the fact that A_1 dominates A_2.

	S_3	S_4
A_1:	I win \$50	*I lose* \$5
A_2:	*I win* \$49	I lose \$6

Let us consider another example: Suppose that I am playing roulette on a rigged wheel, and that the owner of the casino offers me a chance to choose between actions A_1 and A_2 so that the following matrix describes the situation (where S_1=black comes up on the next spin; S_2=red comes up on the next spin):

	S_1	S_2
A_1:	I win \$10	I win \$100
A_2:	I win \$5	I win \$90

Finally suppose that I know that the owner's employee, who is overseeing the wheel and who I am confident is completely loyal to the owner, has been instructed to make black come up on the next spin if I choose A_1 and to make red come up on the next spin if I choose A_2. Clearly even though A_1 dominates A_2, given my knowledge of the situation I should choose A_2. I take it that this needs no argument. It seems that the reason that I should not be guided by dominance considerations is that the states S_1 and S_2 are not probabilistically independent of my actions A_1 and A_2. We can set up the situation so that the states are probabilistically independent of the actions. But when set up in this way, I am led, given my utility assignment to the outcomes, to do A_2.

Let S_3 = the fellow running the roulette wheel follows his boss' instructions; S_4 = the fellow running the roulette wheel disobeys his boss' instructions. (Note that $S_3 = A_1$ & S_1 or A_2 & S_2; $S_4 = A_1$ & S_2 or A_2 & S_1.)

We then have the following matrix:

	S_3	S_4
A_1:	I win \$10	I win \$100
A_2:	I win \$90	I win \$5

Even if I am not sure that S_3 is true, so long as the personal probability of S_3 for me is sufficiently high, I will be led to do A_2, given my utility assignment to the outcomes.

These examples suggest that one should not apply the dominance principle to a situation where the states are not probabilistically independent of the actions. One wishes instead to maximize the expected utility. However, the probabilities that are to be used in determining the expected utility of an action must now be the conditional probabilities of the states given that the action is done. (This is true generally. However when the states are probabilistically independent of the actions, the conditional probability of each state given that one of the actions is done will be equal to the probability of the state, so the latter may be used.) Thus in the roulette wheel example, we may still look at the first matrix given. However, one does not wish to apply the dominance principle but to find the expected utility of the actions, which in our example are:

$$\text{E.U.}(A_1) = \text{prob}(S_1/A_1) \times u(\text{/I win \$10/}) + \text{prob}(S_2/A_1) \times u(\text{/I win \$100/})$$

$$\text{E.U.}(A_2) = \text{prob}(S_1/A_2) \times u(\text{/I win \$5/}) + \text{prob}(S_2/A_2) \times u(\text{/I win \$90/}).$$ [9]

The following position appropriately handles the examples given thus far (ignoring Newcomb's example with which the paper opens) and has intuitive appeal.[10]

(1) It is legitimate to apply dominance principles if and only if the states are probabilistically independent of the actions.

(2) If the states are not probabilistically independent of the actions, then apply the expected utility principle, using as the probability-weights the conditional probabilities of the states given the actions.

Thus in the following matrix, where the entries in the matrix are utility numbers,

	S_1	S_2	....	S_n
A:	O_1	O_2	....	O_n
B:	U_1	U_2	....	U_n

the expected utility of A is $\sum_{i=1}^{n} \text{prob}(S_i/A)\ O_i$, and the expected utility of B is $\sum_{i=1}^{n} \text{prob}(S_i/B)\ U_i$.

III

Is this position satisfactory? Consider the following example: P knows that S or T is his father, but he does not know which one is. S died of some terrible inherited disease, and T did not. It is known that this disease is genetically dominant, and that P's mother did not have it, and that S did not have the recessive gene. If S is his father, P will die of this disease; if T is his father, P will not die of this disease. Furthermore, there is a well-confirmed theory available, let us imagine, about the genetic transmission of the tendency to decide to do acts which form part of an intellectual life. This tendency is genetically dominant. S had this tendency (and did not have the recessive gene), T did not, and P's mother did not. P is now deciding whether (a) to go to graduate school and then teach, or (b) to become a professional baseball player. He prefers (though not enormously) the life of an academic to that of a professional athlete.

	S is P's father	T is P's father
A:	x	y
B:	z	w

$x = P$ is an academic for a while, and then dies of the terrible disease; $z = P$ is a professional athlete for a while, and then dies of the terrible disease; $y = P$ is an academic and leads a normal academic life; $w = P$ is a professional athlete and leads the normal life of a professional athlete, though doing a bit more reading; and P prefers x to z, and y to w. However, the disease is so terrible that P greatly prefers w to x. The matrix might be as follows:

	S is P's father	T is P's father
A:	−20	100
B:	−25	95

Suppose that our well-confirmed theory tells us, and P, that if P chooses the academic life, then it is likely that he has the tendency to choose it; if he does not choose the academic life, then it is likely that he does not have the tendency. Specifically

prob (P has the tendency/P decides to do A) = .9
prob (P does not have the tendency/P decides to do A) = .1
prob (P has the tendency/P decides to do B) = .1
prob (P does not have the tendency/P decides to do B) = .9.

Since P has the tendency iff S is P's father, we have

prob (S is P's father/P decides to do A) = .9
prob (T is P's father/P decides to do A) = .1
prob (S is P's father/P decides to do B) = .1
prob (T is P's father/P decides to do B) = .9.

The dominance principle tells P to do A rather than B. But according to the position we are now considering, in situations in which the states are not probabilistically independent of the actions, the dominance principle is not to be used, but rather one is to use the expected utility principle with the conditional probabilities as the weights. Using the above conditional probabilities and the above numerical assumptions about the utility values, we get:

The expected utility of $A = .9 \times -20 + .1 \times 100 = -8$
The expected utility of $B = .1 \times -25 + .9 \times 95 = 83$.

Since the expected utility of B is greater than that of A, the position we are considering would have P do B rather than A. But this recommendation is perfectly wild. Imagine P saying, 'I am doing B because if I do it it is less likely that I will die of the dread disease'. One wants to reply, 'It is true that you have got the conditional probabilities correct. If you do A it is likely that S is your father, and hence likely that you will die of the disease, and if you do B it is likely that T is your father and hence unlikely that you will die of the disease. But which one of them is your father is already fixed and determined, and has been for a long time. The action you perform legitimately affects our estimate of the probabilities of the two states, but which state obtains does not depend on your action at all. By doing B you are not *making* it less likely that S is your father, and by doing B you are not making it less likely that you will die of the disease'. I do not claim that this reply is without its problems.[11] Before considering another example, let us first state a principle not under attack:

> The Dominance Principle is legitimately applicable to situations in which the states are probabilistically independent of the actions.[12]

If the states are not probabilistically independent of the actions, it *seems* intuitive that the expected utility principle is appropriate, and that it is not legitimate to use the dominance principle if it yields a different result from the expected utility principle. However, in situations in which the states, though not probabilistically independent of the actions, are already fixed and determined, where the actions do not affect whether or not the states obtain, then it *seems* that it is legitimate to use the dominance principle, and illegitimate to follow the recommendation of the expected utility principle if it differs from that of the dominance principle.

For such situations – where the states are not probabilistically independent of the actions, though which one obtains is already fixed and determined – persons may differ over what principle to use.

Of the twelve sorts of situation in which it is not the case both that none of the states are already fixed and determined and none of the states are probabilistically independent of the actions, I shall discuss only one; namely, where each of the states is already fixed and determined, and none of the states are probabilistically independent of the alternative actions.[13]

The question before us is: In this sort of situation, in which all of the states are already fixed and determined, and none of the states are probabilistically independent of the acts, and the dominance principle requires that one do one action, whereas the expected utility principle requires that one do another, should one follow the recommendation of the dominance principle or of the expected utility principle?

The question is difficult. Some may think one should follow the recommendation of the dominance principle; others may think one should follow the recommendation of the expected utility principle in such situations.

Now for the example which introduces a bit of reflexivity which I hope will soon serve us in good stead. Suppose that there are two inherited tendencies ('tendencies' because there is some small probability that it would not be followed in a specific situation):

(1) an inherited tendency to think that the expected utility principle should be used in such situations. (If P has this tendency, he is in state S_1.)

(2) an inherited tendency to think that the dominance principle should be used in such situations. (If P has this tendency, he is in state S_2.)

It is known on the basis of *post mortem* genetic examinations that

(a) P's mother had two neutral genes. (A gene for either tendency genetically dominates a neutral gene. We need not here worry about the progeny who has a gene for each tendency.)

(b) One of the men who may be P's father, had two genes for the first tendency.

(c) The other man who may be P's father had two genes for the second tendency.

So it is known that P has one of the tendencies, but it is not known which one he has. P is faced with the following choice:

	S_1	S_2
A:	10	4
B:	8	3

The choice matrix might have arisen as follows. A deadly disease is going around, and there are two effective vaccines against it. (If both are given, the person dies.) For each person, the side effects of vaccine B are worse than that of vaccine A, and each vaccine has worse side effects on persons in S_2 than either does on persons in S_1.

Now suppose that the theory about the inherited tendencies to choice, tells us, and P knows this, that from a person's choice in *this* situation the probabilities of his having the two tendencies, given that he has one of the two, can be estimated, and in particular

$$\text{prob}\,(S_1/A) = .1$$
$$\text{prob}\,(S_2/A) = .9$$
$$\text{prob}\,(S_1/B) = .9$$
$$\text{prob}\,(S_2/B) = .1.$$

What should P do? What would you do in this situation?

P may reason as follows: if I do A, then very probably S_2 obtains, and I will get 4. If I do B, then very probably S_1 holds, and I will get 8. So I will do B rather than A.

One wants to reply: whether S_1 or S_2 obtains is already fixed and determined. What you decide to do would not bring about one or the other of them. To emphasize this, let us use the past tense. For you are in S_1 iff you were in S_1 yesterday; you are in S_2 iff you were in S_2 yesterday. But to reason 'If I do *A* then very probably I was in S_2 yesterday, and I will get 4. If I do *B*, then very probably, I was in S_1 yesterday, and I will get 8. So I will now do *B* rather than *A*' is absurd. What you decide to do does not affect which state you were in yesterday. For either state, over which you have no control, you are better off doing *A* rather than *B*. To do *B* for reasons such as the above is no less absurd than someone who has already taken vaccine *B* yesterday doing some other act *C* today because the prob (He was in S_1 yesterday/He does *C* today) is very high, and he wants the (delayed) side effects of the vaccine he has already taken to be less severe.

If an explanation runs from *x* to *y*, a correct explanatory theory will speak of the conditional probability prob (y/x). Thus the correct explanatory theory of *P*'s choice in this situation will speak of

prob (*P* does *A*/*P* is in S_1)
prob (*P* does *A*/*P* is in S_2)
prob (*P* does *B*/*P* is in S_1)
prob (*P* does *B*/*P* is in S_2).

From these, the theory may enable us to determine

prob (*P* is in S_1/*P* does *A*)
prob (*P* is in S_2/*P* does *A*)
prob (*P* is in S_1/*P* does *B*)
prob (*P* is in S_2/*P* does *B*)

but these would not be the basic explanatory probabilities. Supposing that probabilistic explanation is legitimate, we could explain why *P* does *A* by having among our antecedent conditions the statement that *P* is in S_2 ,but we cannot *explain* why *P* is in S_2 by having among our antecedent conditions the statement that *P* does *A* (though *P*'s doing *A* may be our reason for believing he is in S_2). Given that when the explanatory line runs from *x* to *y* (*x* is part of the explanation of *y*) and not from *y* to *x*, the theory will speak of and somehow distinguish the conditional probabilities prob (y/x), then the probability prob (x/y) will be a *likeli-*

hood (as, I think, this term is used in the statistical literature). Looking at the likelihoods of the states given the actions may perhaps give one the illusion of control over the states. But I suggest that when the states are already fixed and determined, and the explanatory theory has the influence running from the states to the actions, so that the conditional probabilities of the states on the actions are likelihoods, then if the dominance principle applies, it should be applied.

If a state is part of the explanation of deciding to do an action (if the decision is made) and this state is already fixed and determined, then the decision, which has not yet been made, cannot be part of the explanation of the state's obtaining. So we need not consider the case where prob (state/action) is in the basic explanatory theory, for an already fixed state.[14] What other possibilities are there for already fixed and determined states? One possibility would be a situation in which the states are not part of the explanation of the decision, and the decision is not part of the explanation of which state obtains, but some third thing is part of the explanation of the states obtaining, and the decision's being made. Hence neither prob (state of the matrix obtaining/*P* does a specific action) nor prob (*P* does a specific action/state of the matrix obtains) would be part of the basic explanatory theory (which has conditional probabilities from antecedent to consequent going in the direction of explanation).

Let us consider a case like this, whose matrix exemplifies the structure of the prisoners' dilemma situation, much discussed by game theorists.[15] There are two people, (I) and (II) and the following matrix describes their situation (where the first entry in each box represents the payoff to person (I) and the second entry represents the payoff to person (II)). The situation arises just once, and the persons cannot get together to agree upon a joint plan of action.

		(II)	
		C	*D*
(I)	*A*:	10, 3	4, 4
	B:	8, 8	3, 10

Notice that for person (I), action *A* dominates action *B*, and for person (II), action *D* dominates action *C*. Hence if each performs his dominant action, each ends up with 4. But if each performs the non-dominant action, each ends up with 8. So, in this situation, both persons' following the dominance principle leaves each worse off than if both did not follow the dominance principle.

People may differ over what should be done in this situation. Let us, once again, suppose that there are two inherited tendencies, one to perform the dominant action in this situation, and one to perform the other action. Either tendency is genetically dominant over a possible third inherited trait. Persons (I) and (II) are identical twins, who care only about their own payoffs as represented in this matrix, and knows that their mother had the neutral gene, one of their two possible fathers had only the gene to perform the dominant action, and the other had only the gene not to perform the dominant action. Neither knows which man was their father, nor which of the genes they have. Each knows, given the genetic theory, that it is almost certain that if he performs the dominant (dominated) action his brother will also. We must also suppose that the theory tells us and them that given all this information upon which they base their choice, the correlation between their actions holds as almost certain, and also given *this* additional information, it holds as almost certain, etc.

I do not wish here to discuss whether one should or should not perform the dominant action in Prisoners' Dilemma situations. I wish merely to consider the following argument for not performing the dominant action in the situation I have just described. Suppose brother I argues: 'If I perform the dominant action then it is almost $certain_1$ that I have that gene, and therefore that my brother does also, and so it is almost $certain_2$[16] that he will also perform the dominant action and so it is almost $certain_2$ that I will get 4. Whereas if I perform the dominated action, for similar reasons, it is almost certain that my brother will also, and hence it is almost certain that I will get 8. So I should perform the dominated action'.

Here one surely wishes to reply that *this* argument is not a good argument for performing the dominated action. For what this brother does will not affect what the other brother does. (To emphasize this, suppose that brother II has already acted, though brother I does not yet know what he has done.) Perhaps in prisoners' dilemma situations one should perform the dominated action, but *this* argument does not show that one should in this situation.

The examples thus far considered lead me to believe that if the actions or decisions to do the actions do not affect, help bring about, influence, etc., *which* state obtains, then whatever the conditional probabilities (so long as they do not indicate an influence), one should perform the dominant action.

If the considerations thus far adduced are convincing, then it is clear that one should also choose the dominant action in the following situations, having the same structure (matrix) as Newcomb's, and differing only in that:

(1) The being makes his prediction and sets the process going whereby the $\$M$ gets placed in the second box, or not. You then make your choice, and *after* you do, the (long) process terminates and the $\$M$ gets in the box, or not. So while you are deciding, the $\$M$ is not already there, though at this time he has already decided whether it will be or not.

(2) The being gathers his data on the basis of which he makes his prediction. You make your choice (e.g., press one of two buttons which will open one or both boxes later by delayed action), and he then makes his prediction, on the basis of the data previously gathered, and puts the $\$M$ in, or not.

This suggests that the crucial fact is *not* whether the states are already fixed and determined but whether the actions *influence* or *affect* which state obtains.

Setting up a simple matrix,[17] we have the following possibilities (with the matrix entries being recommended decision policies for the situation).

	A dominant action is available	No dominant action is available
The actions influence which state obtains. The conditional probabilities differ.	(I) Maximize Expected Utility	(II) Maximize Expected Utility
No influence of actions on states. However conditional probabilities differ.	(III)	(IV)
No influence of actions on states. The conditional probabilities are all the same.	(V) Do dominant action (or, equivalently, Maximize Expected Utility)	(VI) Maximize Expected Utility

The standard theories make the recommendations in (V) and (VI).

They do not consider (I) and (II), but (ignoring other difficulties there might be with the policy) Maximizing Expected Utility seems reasonable here. The difficulties come in the middle row. (III) is the situation exemplified by Newcomb's situation and the other examples we have listed from the person choosing whether to lead the academic life, onwards. I have argued that in these situations, one should choose the dominant action and ignore the conditional probabilities which do not indicate an influence. What then should one do in situation (IV), where which action is done does not influence which state obtains, where the conditional probabilities of the states given the actions differ, and where *no* dominant action is available. If the lesson of case (III) is that one should ignore conditional probabilities which do not indicate an influence, must not one ignore them completely in case (IV) as well?

Not exactly. What one should do, in a choice between two actions A and B, is the following.[18] Let $p_1, \ldots, p_n$ be the conditional probability distribution of action A over the n states; let $q_1, \ldots, q_n$ be the conditional probability distribution of action B over the n states. A probability distribution $r_1, \ldots, r_n$, summing to 1, is between $p_1, \ldots, p_n$ and $q_1, \ldots, q_n$ iff for each i, r_i is in the closed interval $[p_i, q_i]$ or $[q_i, p_i]$. (Note that according to this account, $p_1, \ldots, p_n$, and $q_1, \ldots, q_n$ are each between $p_1, \ldots, p_n$ and $q_1, \ldots, q_n$.) Now for a recommendation: If relative to each probability distribution between $p_1, \ldots, p_n$ and $q_1, \ldots, q_n$, action A has a higher expected utility than action B, then do action A. The expected utility of A and B is computed with respect to the same probability distribution. It will not, of course, be the case that relative to every possible probability distribution A has a higher expected utility than B. For, by hypothesis, A does not dominate B. However it may be that relative to each probability distribution between $p_1, \ldots, p_n$ and $q_1, \ldots, q_n$, A has a higher expected utility than B. If, on the other hand, it is not the case that relative to each probability distribution between $p_1, \ldots, p_n$ and $q_1, \ldots, q_n$, A has a higher expected utility than B (and it is not the case that relative to each, B has a higher expected utility than A), then we are faced with a problem of decision under constrained uncertainty (the constraints being the end probability distributions), on which kind of problem there is not, so far as I know, agreement in the literature.[19] Since consideration of the issues raised by such problems would take us far afield, we thankfully leave them.

To talk more objectively than some would like, though more intuitively than we otherwise could, since the actions do not affect or influence which state obtains, there is some one probability distribution, which we do not know, relative to which we would like to compare the action *A* and *B*. Since we do not know the distribution, we cannot proceed as in cases (V) and (VI). But since there is *one* unknown correct distribution 'out there', unaffected by what we do, we must, in the procedure we use, compare each action with respect to the *same* distribution. Thus it is, at this point, an irrelevant fact that one action's expected utility computed with respect to one probability distribution is higher than another action's expected utility computed with respect to *another* probability distribution. It may seem strange that for case (IV) we bring in the probabilities in some way (even though they do not indicate an influence) whereas in case (III) we do not. This difference is only apparent, since we could bring in the probabilities in case (III) in exactly the same way. The reason why we need not do this, and need only note that *A* dominates *B*, is that if *A* dominates *B*, then relative to each probability distribution (and therefore for each one between the conditional ones established by the two actions) *A* has a higher expected utility than *B*.[20]

Now, at last, to return to Newcomb's example of the predictor. If one believes, for this case, that there is backwards causality, that your choice causes the money to be there or not, that it causes him to have made the prediction that he made, then there is no problem. One takes only what is in the second box. Or if one believes that the way the predictor works is by looking into the future; he, in some sense, sees what you are doing, and hence is no more likely to be wrong about what you do than someone else who is standing there at the time and watching you, and would normally see you, say, open only one box, then there is no problem. You take only what is in the second box. But suppose we establish or take as given that there is no backwards causality, that what you actually decide to do does not affect what he did in the past, that what you actually decide to do is not part of the explanation of why he made the prediction he made. So let us agree that the predictor works as follows: He observes you sometime before you are faced with the choice, examines you with complicated apparatus, etc., and then uses his theory to predict on the basis of this state you were in, what choice you would make later when faced with the choice. Your deciding to do as you do is not part of the explanation of why he makes the prediction he does, though your being in a certain state earlier, is part of the explanation of why he makes the prediction he does, and why you decide as you do.

I believe that one should take what is in both boxes. I fear that the considerations I have adduced thus far will not convince those proponents of taking only what is in the second box. Furthermore I suspect that an adequate solution to this problem will go much deeper than I have yet gone or shall go in this paper. So I want to pose one question. I assume that it is clear that in the vaccine example, the person should not be convinced by the probability argument, and should choose the dominant action. I assume also that it is clear that in the case of the two brothers, the brother should not be convinced by the probability argument offered. The question I should like to put to proponents of taking only what is in the second box in Newcomb's example (and hence not performing the dominant action) is: what is the difference between Newcomb's example and the other two examples which make the difference between not following the dominance principle, and following it?

If no such difference is produced, one should not rush to conclude that one should perform the dominant action in Newcomb's example. For it must be granted that, at the very least, it is not *as clear* that one should perform the dominant action in Newcomb's example, as in the other two examples. And one should be wary of attempting to force a decision in an unclear case by producing a similar case where the decision is clear, and challenging one to find a difference between the cases which makes a difference to the decision. For suppose the undecided person, or the proponent of another decision, cannot find such a difference. Does not the forcer, now, have to find a difference between the cases which explains why one is clear, and the other is not? And might not *this* difference then be produced by the other person as that which perhaps should yield different decisions in the two cases? Sometimes this will be implausible; e.g., if the difference is that one case is relatively simple, and the other has much additional detail, individually irrelevant, which prevent the other case from being taken in as a whole. But it does seem that someone arguing as I do about a case must not only: (a) describe a similar case which is clear, and challenge the other to state a difference between them which should make a difference to how they are handled, but must also, (b) describe a difference between the cases which explains why though one case is clear, the other is not, or one is tempted to handle the other case differently. And, assuming that all accept the difference stated in (b) as explaining what it is supposed to explain,

(I) The simplest situation is that in which all agree that the difference mentioned in (b) is not a reason for different decisions in the two cases.

(II) However, if the forcer says it is not a reason for different decisions in the two cases, and the other person says it is or may be, difficult questions arise about upon whom, if anyone, the burden of further argument falls.

What then is the difference that makes some cases clear and Newcomb's example unclear, yet does not make a difference to how the cases should be decided? Given my account of what the crucial factors are (influence, etc.) my answer to this question will have to claim that the clear cases are clear cases of no influence (or, to recall the cases which we considered at the beginning, of influence), and that in Newcomb's example there is the *illusion* of influence. The task is to explain in a sufficiently forceful way what gives rise to this illusion so that, even as we experience it, we will not be deceived by it.

I have said that if the action is referred to in an explanation of the state's obtaining, so that the doing of the action affects or influences which state obtains, then the Dominance Principle should not be applied. And if the explanation of the states' obtaining does not make reference to the action, the action does not influence which state obtains, does not (partly) bring it about that a state obtains, then the Dominance Principle should be applied to such situations where a dominant action is available. But if this is so, where is there room for unclarity about a case? What other possibility is there? Either the action is referred to in the explanation of the state's obtaining, or it is not. How does the temptation to take only what is in the second box arise in the Newcomb example, and why does it linger?

The possibility to which I wish to call attention can be described differently, depending upon other views which one holds. (I describe the possibility specifically with Newcomb's example in mind.) (1) The action *is* referred to in the explanation of the state's obtaining, but the term which refers to the action occurs in the explanation, in a non-extensional belief context. Thus it does not follow from the fact that the action is referred to, in this way, in the explanation of the state's obtaining, that the doing of the action affects which state obtains. (2) The action is not referred to in the explanation of the state's obtaining. What is brought in by the explanation of the state's obtaining is some being's well-founded beliefs about the person's doing the action. Since the person's doing the action is not part of the explanation of the state's obtaining, it does not affect or influence which state obtains.

In Newcomb's example, the predictor makes his prediction on the basis of determining what state the person is in, and what his theory tells him about what such a person will do in the choice situation. Believing his theory accurate, he puts the money in or not, according to his belief about the person's future actions, where this belief depends upon his finding out what initial state the person is in, and what his theory tells him about a person in such an initial state. Thus, if the predictor puts the $\$M$ in the second box, part of the explanation of this is his belief that the person will take only what is in the second box. If he does not put the $\$M$ in the second box, part of the explanation of this is his belief that the person will take what is in both boxes. Thus the explanation of the money's being in the second box (or not) refers to the person's action only in a nonextensional belief context (or does not refer to it at all but only to the predictor's beliefs about it).

It is apparently a persistent temptation for people to believe, when an explanation of something x brings in terms referring to y in a nonextensional belief context (or brings in beliefs about y), that y, in some way, influences or affects x. Thus one finds writers on teleological explanation having to state that in the simple case where someone goes to the refrigerator to get an apple, it is not the apple's being there when he gets there which caused him to go, or which (partly) explains his actions, but rather his beliefs about an apple's being there. But this cannot be the whole story about Newcomb's example. For there are many persons not at all tempted to say that the apple's being there when he gets there influenced his action of going there, who do want to or are tempted to take only what is in the second box.

Let us return to the writers on teleology. To show that the apple's being there does not influence the person's actions, but rather it is his beliefs about the apple's being there that do, they usually argue that even if the apple were not there, so long as the person had the beliefs, he would act in the same way. The relevant feature of nonextensional belief contexts here is that from *P* believes that ...*x*..., it does not follow that *x* exists, from *P* believes that *p*, it does not follow that *p* is true. So, the argument runs, he *could* have his beliefs without there being an apple there, and this shows that the apple does not influence his actions in this case. And surely the explanation of his action should be the same, in the case where the apple is in the refrigerator, as in the case where it is not though he believes it is. The parallel argument for Newcomb's example would run: The predictor could believe that you will take only the second even if you do not. This shows that your action does not influence whether or not the money is there, but rather the predictor's beliefs about your action has this influence. But by the conditions of the problem, the predictor is almost certain to predict correctly, so that it is not clear that the predictor could believe that you will take only the second even it you do not. Thus, the condition of the problem which has the predictor's predictions almost certainly being correct tends to get us to treat the predictor's beliefs as though they do not have these nonextensional features. For if his predictions are almost certainly correct, then almost certainly: if he believes you will do *A* then you will do *A*.

One further thing should be mentioned. It is a reasonably intuitive principle that if *R* brings it about that *p*, and if *p* if and only if *q* (for some 'iff' stronger than the material biconditional), then *R* brings it about that *q*. Or, if it is up to *R* whether *p*, and *p* iff *q* (for some strong 'iff'), then it is up to *R* whether *q*. Thus one finds writers arguing that if there are necessary and sufficient causal conditions for our actions, which conditions go back to a time before we were born, then what we do is not up to us. For, so the argument runs, those conditions obtaining before we were born clearly were not up to us, and so what they are necessary and sufficient for is not up to us either. I do not wish here to discuss whether this principle is correct. All that is required for my purposes is that the principle have intuitive appeal, and be a hard one to escape.

This would also reinforce the feeling that as choosers in Newcomb's example, we can, somehow, influence what the predictor did. For, one might argue, Newcomb's problem is a problem for the chooser only if what he does is up to him. And if one assumes this, and the principle is operating, then it will be difficult to escape the feeling that what the predictor did is up to you, the chooser.

I do not claim that this last principle alone creates the problem. For the problem does not arise in e.g., the vaccine case.[21] But it does, I believe, contribute to it.

Thus I wish to claim that Newcomb's example is less clear than the others because

(a) in it the explanation of the state's obtaining refers to the action (though this reference occurs in a nonextensional belief-context)

and that

(b) the conditions of the problem prevent one obvious way of refuting the teleologist's view, in this case, which view depends upon the truth that generally if y is part of the explanation of x, then y influences x).

This leads to the feeling that, somehow, you as chooser can influence what the predictor did, and this feeling is perhaps reinforced by the operation of the intuitive principle. All this leads to the lurking feeling that one can now choose to take only what is in the second box, and so make oneself the sort of person who does so, and so, somehow, influence what the predictor did. I hope you find this explanation of why some cases are clear and Newcomb's is not, acceptable, and that it is clear that this difference between the cases should not make a difference to how they are decided.[22]

At this point one perhaps wants to say, 'If you produce a case having the features you say distinguish Newcomb's example from the others, where it is clear that the dominant action should be performed, then I will be convinced that the dominant action should be performed in Newcomb's example. But not until'. If I am right about the role of similar examples, then this cannot be done; an answer to Newcomb's example cannot be forced in this way. Or rather, if it can be done, then it will show that I have not picked out the right difference. For if one case that fits my description is clear, and another which fits it is not clear, then we still have to produce features to explain why one is clear and the other is not. And perhaps *those* features should make a difference between the decisions in the two cases. At some point, given an acceptable explanation of why one case is clear and another is not, one just has to see that the explanatory features do not make a difference to what should be decided in the two cases. Or, at any rate, the point that the explanatory features do not make a difference to what should be decided can itself be forced by a clear case only at the cost of the claim that those very features explain why some cases are clear and others are not.

In closing this paper, I must muddy up the waters a bit (more?).

(1) Though Newcomb's example suggests much about when to apply the dominance principle, and when to apply the expected utility principle (and hence is relevant to formal decision theory), it is not the expected utility principle which leads some people to choose only what is in the second box. For suppose the probability of the being's predicting correctly was just .6.

> Then the expected utility of taking what is in both boxes = prob (he predicts correctly /I take both) $\times$ u(I receive \$1000) + prob (he predicts correctly/ I take only second) $\times$ u(I receive \$1001000) = .6 $\times$ u(\$1000) + .4 $\times$ u(\$1001000).

The expected utility of taking only what is in the second box = .6 $\times$ u(\$1000000) + .4 $\times$ u(\$0).

And given the utility I assume each of my readers assigns to obtaining these various monetary amounts, the expected utility of taking only what is in the second box is greater than the expected utility of taking what is in both boxes. Yet, I presume, if the probability of the beings predicting correctly were only .6, each of us would choose to take what is in both boxes.

So it is not (just) the expected utility argument that operates here to create the problem in Newcomb's example. It is crucial that the predictor is almost certain to be correct. I refrain from asking a proponent of taking only what is in the second box in Newcomb's example: if .6 is not a high enough probability to lead you to take only what is in the second box, and almost certainty of correct predictions leads you to take only the second, what is the minimum probability of correct prediction which leads you to take only what is in the second box? I refrain from asking this question because I am very unsure about the force of drawing-the-line arguments, and also because the person who wishes to take what is in both boxes may also face a problem of drawing the line, as we shall see in a moment.

(2) If the fact that it is almost certain that the predictor will be correct is crucial to Newcomb's example, this suggests that we consider the case where it *is* certain, where you know the prediction is correct (though you

do not know what the prediction is. Here one naturally argues: I know that if I take both, I will get \$1000. I know that if I take only what is in the second, I get \$$M$. So, of course, I will take only what is in the second. And does a proponent of taking what is in both boxes in Newcomb's example, (e.g., me) really wish to argue that it is the probability, however, minute, of the predictor's being mistaken which make the difference? Does he really wish to argue that if he knows the prediction will be correct, he will take only the second, but that if he knows someone using the predictor's theory will be wrong once in every 20 billion cases, he will take what is in both boxes? Could the difference between one in n, and none in n, for arbitrarily large finite n, make this difference? And how exactly does the fact that the predictor is certain to have been correct dissolve the force of the dominance argument?

To get the mind to really boggle, consider the following.

	S_1	S_2
A:	10	4
B:	8	3

Suppose that you know that either S_1 or S_2 already obtains, but you do not know which, and you know that S_1 will cause you to do B, and S_2 will cause you to do A. Now choose! ('Choose?')

To connect up again with a causalized version of Newcomb's example, suppose you know that there are two boxes, (B1) and (B2). (B1) contains \$1000. (B2) contains either a valuable diamond or nothing. You have to choose between taking what is in both boxes, and taking only what is in the second. You know that there are two states: S_1 and S_2. You do not know which obtains, but you know that whichever does, it has obtained for the past week. If S_2 obtains, it causes you to take only what is in the second, and it has already caused a diamond to be produced in box (B2). If S_1 obtains, it causes you to take what is in both boxes, and does not cause a diamond to be produced in the second box. You know all this. What do you choose to do?

While we are at it, consider the following case where what you decide (and why) either (1) does affect which future state will obtain, upon which consequences depend, or (though this would not be the same problem for the view I have proposed, it might be for yours) even if it does not affect which state obtains, the conditional probabilities of the states, given what you do and why, differ.

	S_1	S_2
A:	live	die
B:	die	live

(1) Apart from your decisions (if you do not know of this matrix, or know of it and cannot reach a decision), prob $S_1 >$ prob S_2

(2) prob (S_1/do A with (1) as reason) < prob (S_2/do A with (1) as reason)

(3) prob (S_1/do B with (2) as reason) > prob (S_2/do B with (2) as reason)

⋮

even (n) prob (S_1/do A with $n - 1$ as reason) < prob (S_2/do A with $n - 1$ as reason)

odd (n) prob (S_1/do B with $n - 1$ as reason) > prob (S_2/do B with $n - 1$ as reason)

⋮

Also: prob (S_1/you do what you do because indifferent between A and B) > prob (S_2/you do what you do because indifferent between A and B)

prob (S_1 /doing A with all of the above as reason) <
prob (S_2/doing A with all of the above as reason)
and
prob (S_1 /doing B with all of the above as reason) >
prob (S_2/doing B with all of the above as reason).

Finally, where 'all this' refers to all of what is above this place, and reflexively, to the next two, in which it appears:

prob (S_1 /doing A with all this as reason) <
prob (S_2/doing A with all this as reason)
and
prob (S_1 /doing B with all this as reason) >
prob (S_2/doing B with all this as reason).

What do you do?

Harvard University

REFERENCES

* It is not clear that I am entitled to present this paper. For the problem of choice which concerns me was constructed by someone else, and I am not satisfied with my attempts to work through the problem. But since I believe that the problem will interest and intrigue Peter Hempel and his many friends, and since its publication may call forth a solution which will enable me to stop returning, periodically, to it, here it is. It was constructed by a physicist, Dr. William Newcomb, of the Livermore Radiation Laboratories in California. I first heard the problem, in 1963, from his friend Professor Martin David Kruskal of the Princeton University Department of Astrophysical Sciences. I have benefitted from discussions, in 1963, with William Newcomb, Martin David Kruskal, and Paul Benacerraf. Since then, on and off, I have discussed the problem with many other friends whose attempts to grapple with it have encouraged me to publish my own. It is a beautiful problem. I wish it were mine.

[1] If the being predicts that you will consciously randomize your choice, e.g., flip a coin, or decide to do one of the actions if the next object you happen to see is blue, and otherwise do the other action, then he does not put the $\$M$ in the second box.

[2] Try it on your friends or students and see for yourself. Perhaps some psychologists will investigate whether responses to the problem are correlated with some other interesting psychological variable that they know of.

[3] If the questions and problems are handled as I believe they should be, then some of the ensuing discussion would have to be formulated differently. But there is no point to introducing detail extraneous to the central problem of this paper here.

[4] This divergence between the dominance principle and the expected utility principle is pointed out in Robert Nozick. *The Normative Theory of Individual Choice*, unpublished doctoral dissertation, Princeton University, Princeton, 1963, and in Richard Jeffrey, *The Logic of Decision*, McGraw-Hill, New York, 1965.

[5] This is shorthand for: action A is done and state S_{12} obtains or action B is done and state S_1 obtains. The 'or' is the exclusive or.

[6] Note that

$$S_1 = A_1 \mathbin{\&} S_3 \text{ or } A_2 \mathbin{\&} S_4$$
$$S_2 = A_1 \mathbin{\&} S_4 \text{ or } A_2 \mathbin{\&} S_3$$
$$S_3 = A_1 \mathbin{\&} S_1 \text{ or } A_2 \mathbin{\&} S_2$$
$$S_4 = A_1 \mathbin{\&} S_2 \text{ or } A_2 \mathbin{\&} S_1$$

Similarly, the above identities hold for Newcomb's example, with which I began, if one lets

S_1 = The money is in the second box.
S_2 = The money is not in the second box.
S_3 = The being predicts your choice correctly.
S_4 = The being incorrectly predicts your choice.
A_1 = You take only what is in the second box.
A_2 = You take what is in both boxes.

[7] State S is not probabilistically independent of actions A and B if prob (S obtains/A is done) $\neq$ prob (S obtains/B is done).

[8] In Newcomb's predictor example, assuming that 'He predicts correctly' and 'He predicts incorrectly' are each probabilistically independent of my actions, then it is

not the case that 'He puts the money in' and 'He does not put the money in' are each probabilistically independent of my actions.

Usually it will be the case that if the members of the set of exhaustive and exclusive states are each probabilistically independent of the actions A_1 and A_2, then it will not be the case that the states equivalent to our contrived states are each probabilistically independent of both A_1 and A_2. For example, suppose prob (S_1/A_1) = prob (S_1/A_2) = = prob (S_1); prob (S_2/A_2) = prob (S_2/A_1) = prob (S_2). Let:

$$S_3 = A_1 \,\&\, S_1 \text{ or } A_2 \,\&\, S_2$$
$$S_4 = A_1 \,\&\, S_2 \text{ or } A_2 \,\&\, S_1$$

If prob $(S_1) \neq$ prob (S_2), then S_3 and S_4 are not probabilistically independent of A_1 and A_2. For prob (S_3/A_1) = prob (S_1/A_1) = prob (S_1), and prob (S_3/A_2) = prob (S_2/A_2) = prob (S_2). Therefore if prob $(S_1) \neq$ prob (S_2), then prob $(S_3/A_1) \neq$ prob (S_3/A_2). If prob (S_1) = prob (S_2) = 1/2, then the possibility of describing the states as we have will not matter. For if, for example, A_1 can be shifted around so as to dominate A_2, then before the shifting it will have a higher expected utility than A_2. Generally, if the members of the set of exclusive and exhaustive states are probabilistically independent of both A_1 and A_2, then the members of the contrived set of states will be probabilistically independent of both A_1 and A_2 only if the probabilities of the original states which are components of the contrived states are identical. And in this case it will not matter which way one sets up the situation.

[9] Note that this procedure seems to work quite well for situations in which the states are not only not probabilistically independent of the actions, but are not logically independent either. Suppose that a person is asked whether he prefers doing A to doing B, where the outcome of A is /p if S_1 and r if S_2/ and the outcome of B is /q if S_2 and r if S_1/. And suppose that he prefers p to q to r, and that S_1 = I do B, and S_2 = I do A. The person realizes that if he does A, S_2 will be the case and the outcome will be r, and he realizes that if he does B, S_1 will be the case and the outcome will be r. Since the outcome will be r in any case, he is indifferent between doing A and doing B. So let us suppose he flips a coin in order to decide which to do. But given that the coin is fair, it is now the case that the probability of $S_1 = 1/2$ and the probability of $S_2 = 1/2$. If we mechanically started to compute the expected utility of A, and of B, we would find that A has a higher expected utility than does B. For mechanically computing the expected utilities, it would turn out that the expected utility of A = $= 1/2 \times u(p) + 1/2 \times u(r)$, and the expected utility of $B = 1/2 \times u(q) + 1/2 \times u(r)$. If, however, we use the conditional probabilities, then the expected utility of A = prob $(S_1/A) \times u(p)$ + prob $(S_2/A) \times u(r) = 0 \times u(p) + 1 \times u(r) = u(r)$. And the expected utility of B = prob $(S_2/B) \times u(q)$ + prob $(S_1/B) \times u(r) = 0 \times u(q) + 1 \times u(r) = u(r)$. Thus the expected utilities of A and B are equal, as one would wish.

[10] This position was suggested, with some reservations due to Newcomb's example, in Robert Nozick, *The Normative Theory of Individual Choice*, *op. cit.* It was also suggested in Richard Jeffrey, *The Logic of Decision*, *op. cit.*

[11] I should mention, what the reader has no doubt noticed, that the previous *example* is not fully satisfactory. For it seems that preferring the academic life to the athlete's life should be as strong evidence for the tendency as is choosing the academic life. And hence P's choosing the athlete's life, though he prefers the academic life, on expected utility grounds does not seem to make it likely that he does not have the tendency. What the example seems to require is an inherited tendency to decide to do A which is such that (1) The probability of its presence cannot be estimated on the

basis of the person's preferences, but only on the basis of knowing the genetic make-up of his parents, or knowing his actual decisions; and (2) The theory about how the tendency operates yields the result that it is unlikely that it is present if the person decides not to do A in the example-situation, even though he makes this decision on the basis of the stated expected utility grounds. It is not clear how, for this example, the details are to be coherently worked out.

[12] That is, the Dominance Principle is legitimately applicable to situations in which $\sim (\exists S)(\exists A)(\exists B)$ [prob (S obtains/A is done) $\neq$ prob (S obtains/B is done)].

[13] The other eleven possibilities about the states are:

	Already fixed and determined		Not already fixed and determined	
	probabilistically independent of the actions	not probabilistically independent of the actions	prob. ind. of the actions	not prob. ind. of the actions
(1)	some	some	some	some
(2)	some	some	some	none
(3)	some	some	none	some
(4)	some	some	none	none
(5)	some	none	some	some
(6)	some	none	some	none
(7)	some	none	none	some
(8)	all	none	none	none
(9)	none	some	some	some
(10)	none	some	some	none
(11)	none	some	none	some

[14] Unless it is possible that there be causality or influence backwards in time. I shall not here consider this possibility, though it may be that only on its basis can one defend, for some choice situations, the refusal to use the dominance principle. I try to explain later why, for some situations, even if one grants that there is no influence back in time, one may not escape the feeling that, somehow, there is.

[15] Cf. R. Duncan Luce and Howard Raiffa, *Games and Decisions*, John Wiley & Sons, New York, 1957, pp. 94–102.

[16] Almost certainty$_1 >$ almost certainty$_2$, since almost certainty$_2$ is some function of the probability that brother I has the dominant action gene given that he performs the dominant action (= almost certainty$_1$), and of the probability that brother II does the dominant action given that he has the dominant action gene.

[17] In choosing the headings for the rows, I have ignored more complicated possibilities, which must be investigated for a fuller theory, e.g., some actions influence which state obtains and others do not.

[18] I here consider only the case of two actions. Obvious and messy problems for the kind of policy about to be proposed are raised by the situation in which more than two actions are available (e.g., under what conditions do pairwise comparisons lead to a linear order), whose consideration is best postponed for another occasion.

[19] See R. Duncan Luce and Howard Raiffa, *op. cit.*, pp. 275–298 and the references therein; Daniel Ellsberg, 'Risk, Ambiguity, and the Savage Axioms', *Quarterly Journal of Economics* **75** (1961), 643–669, and the articles by his fellow symposiasts Howard Raiffa and William Feller.

[20] If the distinctions I have drawn are correct, then some of the existing literature is in need of revision. Many of the writers might be willing to just draw the distinctions we have adumbrated. But for the specific theories offered by some personal probability theorists, it is not clear how this is to be done. For example, L. J. Savage in *The Foundations of Statistics*, John Wiley & Sons, New York, 1954, recommends unrestricted use of dominance principles (his postulate *P*2), which would not do in case (I). And Savage seems explicitly to wish to deny himself the means of distinguishing case (I) from the others. (For further discussion, some of which must be revised in the light of this paper, of Savage's important and ingenious work, see Robert Nozick, *op. cit.*, Chapter V.) And Richard Jeffrey, *The Logic of Decision*, *op. cit.*, recommends universal use of maximizing expected utility relative to the conditional probabilities of the states given the actions (see footnote 10 above). This will not do, I have argued, in cases (III) and (IV). But Jeffrey also sees it as a special virtue of this theory that it does not utilize certain notions, and these notions look like they might well be required to draw the distinctions between the different kinds of cases.

While on the subject of how to distinguish the cases, let me (be the first to) say that I have used without explanation, and in this paper often interchangeably, the notions of influency, affecting, etc. I have felt free to use them without paying them much attention because even such unreflective use serves to open a whole area of concern. A detailed consideration of the different possible cases with many actions, some influencing, and in different degrees, some not influencing, combined with an attempt to state detailed principles using precise 'influence' notions undoubtedly would bring forth many intricate and difficult problems. These would show, I think, that my quick general statements about influence and what distinguishes the cases, are not, strictly speaking, correct. But going into these details would necessitate going into these details. So I will not.

[21] Though perhaps it explains why I *momentarily* felt I had succeeded too well in constructing the vaccine case, and that perhaps one *should* perform the non-dominant action there.

[22] But it also seems relevant that in Newcomb's example not only is the action referred to in the explanation of which state obtains (though in a nonextensional belief context), but also there is another explanatory tie between the action and the state; namely, that both the state's obtaining, and your actually performing the action are both partly explained in terms of some third thing (your being in a certain initial state earlier). A fuller investigation would have to pursue yet more complicated examples which incorporated this.

ADOLF GRÜNBAUM

THE MEANING OF TIME*

I. INTRODUCTION

Studies of time by scientists have often been concerned with the multi-faceted problems of measuring time intervals in atomic, geophysical, biological, and astronomical contexts. It has been claimed that in addition to exhibiting measurable intervals, time is characterized by a *transiency* of the present, which has often been called 'flux' or 'passage'.

Indeed, it has been maintained that '*the passage of time* ... is the very essence of the concept'.[1] I therefore wish to focus my concern with the meaning of time on the credentials which this transiency of the present can claim from the point of view of current physical theories.

In the common-sense view of the world, it is of the very essence of time that events occur now, or are past, or future. Furthermore, events are held to change with respect to belonging to the future or the present. Our commonplace use of tenses codifies our experience that any particular present is superseded by another whose event-content thereby 'comes into being'. It is this occurring *now* or coming into being of previously future events and their subsequent belonging to the past which is called 'becoming' or 'passage'. Thus, by involving reference to *present* occurrence, becoming involves more than mere occurrence at various serially ordered clock times. The past and the future can be characterized as respectively before and after the present. Hence I shall center my account of becoming on the status of the present or now as an attribute of events which is encountered in *perceptual* awareness.

II. THE ISSUE OF THE MIND-DEPENDENCE OF BECOMING

Granted that becoming is a prominent feature of our temporal awareness, I ask: *must* becoming therefore also be a feature of the order of physical events *independently* of our awareness of them, as the common-sense view

N. Rescher et al. (eds.), Essays in Honor of Carl G. Hempel.

supposes it to be? And if not, is there anything within physical theory *per se* to warrant this common-sense conclusion?

It is apparent that the becoming of physical events in our temporal awareness does not itself guarantee that becoming has a mind-independent physical status. Common-sense color attributes, for example, surely *appear* to be properties of physical objects independently of our awareness of them and are held to be such by common sense. And yet scientific theory tells us that they are mind-dependent qualities like sweet and sour are. Of course, if physical theory claims that, contrary to common sense, becoming is not a feature of the temporal order of physical events with respect to earlier and later, then a more comprehensive scientific and philosophical theory must take suitable cognizance of becoming as a conspicuous characteristic of our *temporal awareness* of both physical and mental events.

In this lecture, I aim to clarify the status of temporal becoming by dealing with each of the questions I posed. Clearly, an account of becoming which provides answers to these questions is *not* an analysis of what the common-sense man actually *means* when he says that a physical event belongs to the present, past, or future; instead, such an account sets forth how these ascriptions ought to be construed within the framework of a theory which would supplant the scientifically untutored view of common sense. That the common-sense view is indeed scientifically untutored is evident from the fact that *at a time t*, both of the following physical events qualify as occurring 'now' or 'belonging to the present' according to that view: (i) a stellar explosion that occurred several million years before time t but which is first seen on earth at time t, (ii) a lightning flash originating only a fraction of a second before t and observed at time t. If it be objected that present-day common-sense beliefs have *begun* to allow for the finitude of the speed of light, then I reply that they err at least to the extent of associating absolute simultaneity with the now.

The temporal relations of earlier (before) and later (after) can obtain between two physical events independently of the transient now and of any minds. On the other hand, the classification of events into past, present, and future, which is inherent to becoming, requires reference to the adverbial attribute now as well as to the relations of earlier and later. Hence the issue of the mind-dependence of becoming turns on the status of the adverbial attribute now. And to assert in this context that becoming is mind-dependent is *not* to assert that the obtaining of the relation of temporal precedence among physical events is mind-dependent. Nor is it to assert that the mere occurrence of events at various serially-ordered clock times is mind-dependent.

With these explicit understandings, I can state my thesis as follows: Becoming is mind-dependent because it is not an attribute of physical events per se but requires the occurrence of certain *conceptualized conscious experiences* of the occurrence of physical events. The doctrine that becoming is mind-dependent has been misnamed 'the theory of the block universe'. I shall therefore wish to dissociate the tenets of this doctrine both from serious misunderstandings by its critics and from the very misleading suggestions of the metaphors used by some of its exponents. Besides stating my positive reasons for asserting the mind-dependence of becoming, I shall defend this claim against the major objections which have been raised against it.

III. THE DISTINCTION BETWEEN TEMPORAL BECOMING AND THE ANISOTROPY OF TIME

In order to treat these various issues without risking serious confusions, we must sharply distinguish the following two questions: (i) do physical events *become* independently of any conceptualized awareness of their occurrence, and (ii) are there any kinds of physical or biological processes which are *irreversible* on the strength of the laws of nature and/or of *de facto* prevailing boundary conditions? I shall first state how these two questions have come to be identified and will then explain why it is indeed an error of consequence to identify them. The second of these questions, which pertains to irreversibility, is often formulated by asking whether the time of physics and biology has an 'arrow'. But this formulation of question (ii) can mislead by inviting misidentification of (ii) with (i). For the existence of an arrow is then misleadingly spoken of as constituting a 'one-way forward flow of time', but so also is becoming on the strength of being conceived as the forward 'movement' of the present. And this misidentification is then used to buttress the false belief that an affirmative answer to the question about irreversibility entails an affirmative answer to the question about becoming. To see why I claim that there is indeed a weighty misidentification here, let us first specify what is involved logically when we inquire into the existence of kinds of processes in nature which are irreversible.

If the system of world lines, each of which represents the career of a physical object, is to exhibit a one-dimensional temporal order, relations of simultaneity between spatially separated events are required to define world states. For our purposes, it will suffice to use the simultaneity criterion of some one local inertial frame of the special theory of relativity instead of resorting to the cosmic time of some cosmological model.

Assume now that the events belonging to *each* world-line are invariantly ordered with respect to all inertial systems by a *betweenness* relation having the following *formal* property of the spatial betweenness of the points on an Euclidean straight line: of any three elements, only one can be between the other two. This betweenness of the events is clearly temporal rather than spatial, since it *invariantly* relates the events belonging to each *individual world line* with respect to all inertial systems, while no such *spatial* betweenness obtains invariantly.[2] So long as the temporal betweenness of the world lines is formally Euclidean in the specified sense, any two events on one of them or any two world states can serve to define two time senses which are *ordinally* opposite to each other with respect to the assumed temporal betweenness relations.[3] And the members of the simultaneity-classes of events constituting one of these two opposite senses can then bear lower real number coordinates while those of the other sense can bear the higher coordinates. It is immaterial at this stage which of the two opposite senses is assigned the higher real numbers. All we require is that the real number coordinatization reflect the temporal betweenness relations among the events as follows: events which are temporally between two given events E and E' must bear real number coordinates which are numerically between the time coordinates of E and E'. Employing some one time coordinatization meeting this minimal requirement, we can use the locutions 'initial state', 'final state', 'before', and 'after' on the basis of the magnitudes of the real number coordinates, entirely without prejudice to whether there are irreversible kinds of processes.[4] By an 'irreversible process' (à la Planck) we understand a process such that no counter-process is capable of restoring the original *kind* of state of the system at another time. Note that the temporal vocabulary used in this definition of what is *meant* by an irreversible kind of process does *not* assume tacitly that there *are* irreversible processes: as used here, the terms 'original state', 'restore', and 'counter-process' presuppose only the coordinatization based on the assumed betweenness.

It has been charged that one is guilty of an illicit spatialization of time if one speaks of temporal betweenness while still leaving it *open* whether there are irreversible kinds of processes. But this charge overlooks that the *formal* property of the betweenness on the Euclidean line which I invoked is abstract and, as such, neither spatial nor temporal. And the meaningful attribution of this formal property to the betweenness relation among the events belonging to each world line without any assumption of irreversibility is therefore *not* any kind of illicit spatialization of time. As well say that since temporal betweenness does have this abstract property, the ascription of the latter to the betweenness among the points on a line of space is a temporalization of space![5]

Thus the assumption that the events belonging to each world line are invariantly ordered by an abstractly Euclidean relation of temporal betweenness does not entail the existence of irreversible kinds of processes, but allows every kind of process to be reversible.[6] If there are irreversible processes, then the two ordinally opposite time senses are indeed *further* distinguished structurally as follows: there are certain kinds of sequences of states of systems specified in the order of increasing time coordinates such that these same kinds of sequences do *not* likewise exist in the order of decreasing time coordinates. Or, equivalently, the existence of irreversible processes *structurally* distinguishes the two opposite time senses as follows: there are certain kinds of sequences of states of systems specified in the order of *decreasing* time coordinates such that these same kinds of sequences do *not* likewise obtain in the order of increasing time coordinates. Accordingly, if there are irreversible kinds of processes, then time is *anisotropic*.[7] When physicists say with Eddington that time has an 'arrow', it is this anisotropy to which they are referring metaphorically. Specifically, the spatial opposition between the head and the tail of the arrow represents the structural anisotropy of time.

Note that we were able to characterize a process as irreversible and time as anisotropic without any explicit or tacit reliance on the transient now or on tenses of past, present, and future.[8] By the same token, we are able to assert metaphorically that time has an 'arrow' without any covert or outright reference to events as occurring *now*, happening at present, or coming into being. Nonetheless, the anisotropy of time symbolized by the arrow has been falsely equated in the literature with the transiency of the now or becoming of events via the following steps of reasoning: (1) the becoming of events is described by the kinematic metaphor 'the flow of time' and is conceived as a *shifting* of the now which *singles out the future direction of time* as the sense of its 'advance', and (2) although the physicist's arrow does not involve the transient now, his assertion that there is an arrow of time is taken to be equivalent to the claim that there is a *flow* of time in the direction of the future; this is done by attending to the head of the arrow *to the neglect of its tail* and identifying the former with the direction of 'advance' of the now. The physicist's assertion that time has an 'arrow' discerningly codifies the empirical fact that the two ordinally opposite time senses are *structurally different* in specified respects. But in thus codifying this empirical fact, the physicist does *not* invoke the transient now to single out one of the two time senses as preferred over the other. By contrast, the claim that the present or now shifts in the direction of the future does invoke the transient now to single out one of the two time senses and – as we are about to see – is a mere truism like 'All bachelors are males'. Specifically, the terms 'shift' or 'flow' are used in their literal kinematic senses in such a way that the *spatial* direction of a shift or flow is specified by where the shifting object is at *later* times. Hence when we speak metaphorically of the now as 'shifting' temporally in a particular *temporal* direction, it is then simply a matter of definition that the now shifts or advances in the direction of the future. For this declaration tells us no more than that the nows corresponding to later times are later than those corresponding to earlier ones, which is just as uninformative as the truism that the earlier nows precede the later ones.[9]

It is now apparent that to assert the existence of irreversible processes in the sense of physical theory by means of the metaphor of the arrow does not entail at all that there is a mind-independent becoming of physical events as such. Hence those wishing to assert that becoming is independent of mind cannot rest this claim on the anisotropy of physical time.

Being only a tautology, the kinematic metaphor of time flowing in the direction of the future does not itself render any empirical fact about the time of our experience. But the role played by the present in becoming is a feature of the experienced world codified by common-sense time in the following informative sense: to each of a great diversity of events which are ordered with respect to earlier and later by physical clocks, there corresponds a particular experience of the event as occurring *now*. Hence we shall say that our experience exhibits a *diversity of 'now-contents'* of awareness which are temporally ordered with respect to each of the relations earlier and later. Thus, it is a significant feature of the experienced world codified by common-sense time that there is a sheer diversity of nows, and in that sheer diversity the role of the future is no greater than that of the past. In this *directionally-neutral* sense, therefore, it is informative to say that there is a *transiency* of the now or a coming-into-being of different events. And, of course, in the context of the respective relations of earlier and later, this flux of the present makes for events being past and future.

In order to deal with the issue of the mind-dependence of becoming, I wish to forestall misunderstandings that can arise from uses of the terms 'become', and 'come into being' in senses which are *tenseless*. These senses do not involve belonging to the present or occurring now as understood in tensed discourse, and I must emphasize strongly that my thesis of the mind-dependence of becoming pertains only to the *tensed* variety of becoming. Examples of tenseless uses of the terms 'come into being', 'become', and 'now' are the following: (1) A child *comes into being* as a legal entity the moment it is conceived biologically. What is meant by this incidentally false assertion is that for legal purposes, the career of a child *begins* (tenselessly) at the moment at which the ovum is (tenselessly) fertilized. (2) If gunpowder is suitably ignited at any particular time *t*, an explosion *comes into being* at that time *t*. The species of coming into being meant here involves a common-sense event which is here asserted to *occur tenselessly at time t*. (3) When heated to a suitable temperature, a piece of iron *becomes* red. Clearly, this sentence asserts that after a piece of iron is (tenselessly) suitably heated, it is (tenselessly) red for an unspecified time interval. (4) In Minkowski's two-dimensional spatial representation of the space-time of special relativity, the event shown by the origin-point is called the 'Here-NOW', and correlatively certain event classes in the diagram are respectively called 'Absolute

PAST' and 'Absolute FUTURE'; but Minkowski's 'Here-NOW' denotes an arbitrarily chosen event of reference which can be chosen *once and for all* and continues to qualify as 'now' at various times independently of when the diagram is used. Hence there is no transiency of the now in the relativistic scheme depicted by Minkowski, and his absolute past and absolute future are simply absolutely earlier and absolutely later than the arbitrarily chosen fixed reference event called 'Here-NOW'.[10] Accordingly, we must be mindful that there are tenseless senses of the words 'becoming' and 'now'.

But conversely, we must realize that some important *seemingly* tenseless uses of the terms 'to exist', 'to occur', 'to be actual', and 'to have being or reality' are in fact laden with the present tense. Specifically, all of these terms are often used in the sense of to occur NOW. And by tacitly making the *nowness* of an event a necessary condition for its occurrence, existence, or reality, philosophers have argued fallaciously as follows. They first assert that the universe can be held to exist only to the extent that there are present events. Note that this either asserts that only present events exist now (which is trivial) or it is false. They then invoke the correct premiss that the existence of the physical universe is not mind-dependent and conclude (from the first assertion) that being present, occurring now, or becoming is *independent* of mind or awareness. Thus, Thomas Hobbes wrote: 'The present only has a being in nature; things past have a being in the memory only, but things to come have no being at all, the future being but a fiction of the mind....'[11] When declaring here that only present events or present memories of past events 'have being', Hobbes *appears* to be appealing to a sense of 'to have being' or of 'to exist' which is *logically independent* of the concept of existing-NOW. But his claim depends for its plausibility on the tacit invocation of *present* occurrence as a necessary condition for having being or existing. Once this fact is recognized, his claim that 'the present only has a being in nature' is seen to be the mere tautology that 'only what exists now does indeed exist now'. And by his covert appeal to the irresistible conviction carried by this triviality, he makes plausible the utterly unfounded conclusion that nature can be held to exist only to the extent that there are *present* events and *present* memories of past events. Clearly the fact that an event does not occur now does not justify the conclusion that it does not occur at some time or other.

IV. THE MIND-DEPENDENCE OF BECOMING

Being cognizant of these logical pitfalls, we can turn to the following important question: if a physical event occurs *now* (at present, in the present), what attribute or relation of its occurrence can warrantedly be held to qualify it as such?

In asking this question, I am being mindful of the following fact: if at a given clock time t_0 it is true to say of a particular event E that it is occurring now or happening at present, then this claim could not also be truly made at all other clock times $t \neq t_0$. And hence we must distinguish the tensed assertion of *present* occurrence from the tenseless assertion that the event E occurs at the time t_0: namely, the latter tenseless assertion, if true at all, can truly be made at all times t other than t_0 no less than at the time t_0. By the same token we must guard against identifying the tensed assertion, made at some particular time t_0, that the event E happens *at present* with the tenseless assertion made at *any* time t, that the event E occurs or '*is* present' at time t_0. And similarly for the distinction between the tensed senses of being past or being future, on the one hand, and the tenseless senses of being *past at time* t_0 or being future at time t_0, on the other. To be future at time t_0 just means to be later than t_0, which is a tenseless relation. Thus our question is: what *over and above its otherwise tenseless occurrence at a certain clock time* t, in fact at a time t characterizes a physical event as *now* or as belonging to the present? It will be well remembered from the *Introduction* why my construal of this question does *not* call for an analysis of the common-sense meaning of 'now' or of 'belonging to the present' but for a critical assessment of the status which common sense attributes to the present.[12] Given this construal of the question, my reply to it is: what qualifies a physical event at a time t as belonging to the present or as now is *not* some physical attribute of the event or some relation it sustains to other *purely physical* events. Instead what is *necessary* so to qualify the event is that at the time t at least one human or other *mind-possessing* organism M is conceptually aware of experiencing the event at that time.[13] And that awareness does not, in general, comprise information concerning the date and numerical clock time of the occurrence of the event. What then is the content of M's conceptual awareness at time t that he *is experiencing* a certain event *at that time*? M's experience of the event at time *t* is coupled with an awareness of the temporal coincidence of his experience of the event with a state of *knowing* that he has that experience at all. In other words, *M* experiences the event at *t and* knows that he is experiencing it. Thus, presentness or nowness of an event requires conceptual awareness of the presentational immediacy of the experience of the event. For example, if I just hear a noise at a time *t*, then the noise does not qualify at *t* as *now* unless at *t* I am judgmentally aware of the fact of my hearing it at all and of the temporal coincidence of the hearing with that awareness.[14] If the event at the time *t* is itself a mental event (e.g., a pain), then there is no distinction between the event and our experience of it. With this understanding, I claim that the nowness at a time *t* of either a physical or a mental event requires that there be an *experience* of the event which satisfies the specified requirements. And by satisfying these requirements, the *experience* of a physical event qualifies at the time *t* as occurring *now*. Thus, the fulfillment of the stated requirements by the *experience* of an event at time *t* is also *sufficient* for the nowness of that *experience* at the time *t*. But the mere fact that the experience of a physical event qualifies as now at a clock time *t* allows that in point of physical fact the physical event itself occurred millions of years before *t*, as in the case of now seeing an explosion of a star millions of light years away. Hence, the mere presentness of the experience of a physical event at a time *t* does *not* warrant the conclusion that the clock time of the event is *t* or some *particular* time before *t*. Indeed, the occurrence of an external physical event *E* can never be simultaneous in any inertial system with the direct perceptual registration of *E* by a conceptualizing organism. Hence if *E* is presently experienced as happening at some particular clock time *t*, then there is no inertial system in which *E* occurs at that *same* clock time *t*. Of course, for *some* practical purposes of daily life, a nearby terrestrial flash in the sky can be held to be simultaneous with someone's experience of it with impunity, whereas the remote stellar explosion of a supernova or an eclipse of the sun, for example, may not. But this kind of practical impunity of common-sense perceptual judgments of the presentness of physical events cannot detract from their scientific falsity. And hence I do not regard it as incumbent upon myself to furnish a philosophical account of the status of nowness which is compatible with the now-verdicts of common sense. In particular, I would scarcely countenance making the nowness of the *experience* of a physical event *sufficient* for the nowness of the event, in addition to being necessary, and even informed common sense might balk at this in cases such as a stellar explosion. But all that is essential to my thesis of mind-dependence is that the nowness of the experience of an event is *necessary* for the nowness of the event itself, subject to one qualification to be stated under (3) (b) below. And hence my thesis would allow a compromise with common sense to the following extent: allowing ascriptions of nowness to those physical events which have the very vague relational property of occurring only 'slightly earlier' than someone's appropriate experience of them.

Note several crucial commentaries on my characterization of the now:

(1) My characterization of *present* happening or occurring *now* is intended to *deny* that belonging to the present is a physical attribute of a physical event *E* which is *independent* of any judgmental awareness of the occurrence of *E*. But I am *not* offering any kind of *definition* of the adverbial attribute now, which belongs to the conceptual framework of tensed discourse, solely in terms of attributes and relations drawn from the tenseless (Minkowskian) framework of temporal discourse familiar from physics. In particular, I avowedly invoked the present tense when I made the nowness of an event *E* at time *t* dependent on someone's knowing at *t* that he *is experiencing E*. And this is tantamount to someone's judging at *t*: I am experiencing *E now*. But this formulation is *non*-viciously circular. For it serves to articulate the mind-dependence of nowness, *not* to claim erroneously that it has been eliminated by explicit definition. In fact, I am very much less concerned with the adequacy of the specifics of my characterization than with its thesis of mind-dependence.

(2) It makes the nowness of an event at time *t* depend on the existence of conceptualized awareness that an experience of the event is being had at *t*, and points out the insufficiency of the mere having of the experience. Suppose that at time *t* I express such conceptualized awareness in a linguistic utterance, the utterance being quasi-simultaneous with the experience of the event. Then the utterance satisfies the condition necessary for the *present* occurrence of the experienced event.[15]

(3) *In the first instance*, it is only an experience (i.e., a mental event) which can ever qualify as occurring now, and moreover a mental event (e.g., a pain) must meet the specified awareness requirements in order to qualify. A *physical* event like an explosion can qualify as now at some

time *t* only *derivatively* in one of the following two ways: (a) it is necessary that someone's *experience* of the physical event does so qualify, or (b) if unperceived, the physical event must be simultaneous with another physical event that does so qualify in the derivative sense indicated under (a). For the sake of brevity, I shall refer to this complex state of affairs by saying that physical events belonging to regions of space-time wholly devoid of conceptualizing percipients at no time qualify as occurring now and hence as such do not become.

(4) My characterization of the now is narrow enough to exclude past and future events: It is to be understood here that the *reliving* or anticipation of an event, however vivid it may be, is *not* to be misleadingly called 'having an experience' of the event when my characterization of the now is applied to an experience.

My claim that nowness is mind-dependent does not assert at all that the nowness of an event is arbitrary. On the contrary, it follows from my account that it is not at all arbitrary what event or events qualify as being *now* at any given time *t*: to this extent, my account accords with common sense. But I repudiate much of what common sense conceives to be the status of the now. Thus, when I wonder in thought (which I *may* convey by means of an interrogative verbal utterance) whether it is 3 P.M. Eastern Standard Time now, I am asking myself the following: Is the particular percept of which I am now aware when asking this question a member of the simultaneity class of events which qualify as occurring at 3 P.M., E.S.T. on this particular day? And when I wonder in thought about what is happening now, I am asking the question: What events of which I am not aware are simultaneous with the particular now-percept of which I *am* aware upon asking this question?

That the nowness attribute of an occurrence, when ascribed non-arbitrarily to an event, is inherently mind-dependent seems to me to emerge from a consideration of the kind of information which the judgment 'It is 3 P.M., E.S.T. now' can be warrantedly held to convey. Clearly such a judgment is informative, unlike the judgment 'All bachelors are males'. But if the word 'now' in the informative temporal judgment does not involve reference to a particular content of conceptualized awareness or to the linguistic utterance which renders it at the time, then there would seem to be nothing left for it to designate other than either the time of the events already identified as occurring at 3 P.M., E.S.T. or the time of those identified as occurring at some other time. In the former case, the initially informative temporal judgment 'It is 3 P.M., E.S.T. now' turns into the utter triviality that the events of 3 P.M., E.S.T. occur at 3 P.M., E.S.T.! And in the latter case, the initially informative judgment, if false in point of fact, becomes self-contradictory like '*No* bachelors are males'.

What of the retort to this objection that independently of being perceived physical events themselves possess an unanalyzable property of nowness (i.e., presentness) at their respective times of occurrence over and above merely occurring at these clock times? I find this retort wholly unavailing for several reasons as follows: (1) It must construe the assertion 'It is 3 P.M., E.S.T. now' as claiming *non-trivially* that when the clock strikes 3 P.M. on the day in question, this clock event and all of the events simultaneous with it intrinsically have the unanalyzable property of nowness or presentness. But I am totally at a loss to see that anything non-trivial can possibly be asserted by the claim that at 3 P.M. nowness (presentness) inheres in the events of 3 P.M. For all I am able to discern here is that the events of 3 P.M. are indeed those of 3 P.M. on the day in question! (2) It seems to me of decisive significance that nowness, in the sense associated with becoming, plays no role as a property of physical events themselves in any of the extant theories of physics. There have been allegations in the literature (most recently in H. A. C. Dobbs, 'The "Present" in Physics', *British Journal for the Philosophy of Science* **19** (1968–1969), 317–24) that such branches of statistical physics as meteorology and indeterministic quantum mechanics implicitly assert the existence of a physical counterpart to the human sense of the present. But both below (§V) and elsewhere (in my Reply to Dobbs in the *British Journal for the Philosophy of Science*, August, 1969), I argue that these allegations are mistaken. Hence I maintain that if nowness were a mind-*in*dependent property of physical events themselves, it would be very strange indeed that it could be omitted *as such* from all extant physical theories *without detriment to their explanatory success*. And I hold with Reichenbach[16] that 'if there is Becoming [independently of awareness] the physicist must know it'. (3) As we shall have occasion to note near the end of Section V, the thesis that nowness is *not* mind-dependent poses a serious perplexity pointed out by J. J. C. Smart, and the defenders of the thesis have not even been able to hint how they might resolve that perplexity without utterly trivializing their thesis.

The claim that an event can be now (present) only upon being experienced accords fully, of course, with the common-sense view that there is no more than one time at which a particular event is present and that this time cannot be chosen arbitrarily. But if an event is ever experienced at all such that there is simultaneous awareness of the fact of that experience, then there exists a time at which the event does qualify as being now provided that the event occurs only 'slightly earlier' than the experience of it.

The relation of the conception of becoming espoused here to that of common sense may be likened to the relation of relativity physics to Newtonian physics. My account of nowness as mind-dependent disavows rather than vindicates the common-sense view of its status. Similarly, relativity physics entails the falsity of the results of its predecessor. Though Newtonian physics thus cannot be reduced to relativity physics (in the technical sense of reducing one theory to another), the latter enables us to see why the former works as well as it does in the domain of low velocities: relativity theory shows (via a comparison of the Lorentz and Galilean transformations) that the observational results of the Newtonian theory in that domain are sufficiently correct numerically for some practical purposes. In an analogous manner, my account of nowness enables us to see why the common-sense concept of becoming can function as it does in serving the pragmatic needs of daily life.

A *now-content* of awareness can comprise awareness that one event is later than or succeeds another, as in the following examples: (1) When I perceive the 'tick-tock' of a clock, the 'tick' is not yet part of my past when I hear the 'tock'.[17] As William James and Hans Driesch have noted, melody awareness is another such case of quasi-instantaneous awareness of succession.[18] (2) Memory states are contained in now-contents when we have awareness of other events as being earlier than the event of our awareness of them. (3) A now-content can comprise an envisionment of an event as being later than its ideational anticipation.

V. CRITIQUE OF OBJECTIONS TO THE MIND-DEPENDENCE OF BECOMING

Before dealing with some interesting objections to the thesis of the mind-dependence of becoming, I wish to dispose of some of the caricatures of that thesis with which the literature has been rife under the misnomer of 'the theory of the block universe'. The worst of these is the allegation that the thesis asserts the timelessness of the universe and espouses, in M. Capek's words, the 'preposterous view... that... time is merely a huge and chronic [sic!] hallucination of the human mind'.[19] But even the most misleading of the spatial metaphors that have been used by the defenders of the mind-dependence thesis do not warrant the inference that the thesis denies the objectivity of the so-called 'time-like separation' of events known from the theory of relativity. To assert that nowness, and thereby, pastness and futurity are mind-dependent is surely *not* to assert that the earlier-later relations between the events of a world line are mind-dependent, let alone hallucinatory.

The mind-dependence thesis does deny that physical events themselves happen in the tensed sense of coming into being apart from anyone's awareness of them. But this thesis clearly avows that physical events do happen independently of any mind in the tenseless sense of merely occurring at certain clock times in the context of objective relations of earlier and later. Thus it is a travesty to equate the objective *becominglessness* of physical events asserted by the thesis with a claim of *timelessness*. In this way the thesis of mind-dependence is misrepresented as entailing that all events happen simultaneously or form a '*totum simul*'.[20] But it is an egregious blunder to think that if the time of physics lacks *passage* in the sense of there not being a transient now, then physical events cannot be temporally separated but must all be simultaneous.

A typical example of such a misconstrual of Weyl's and Einstein's denial of physical passage is given by supposing them to have claimed 'that the world is like a film strip: the photographs *are already there* and are merely being exhibited to us'.[21] But when photographs of a film strip 'are already there', they all exist now and hence *simultaneously*. Therefore it is wrong to identify Weyl's denial of physical becoming with the pseudo-image of the 'block universe' and then to charge his denial with entailing the absurdity that all events are simultaneous. Thus Whitrow says erroneously: 'the theory of "the block universe" ... implies that past (and future) events co-exist with those that are present'.[22] We shall see in Section VI that a corresponding error vitiates the allegation that determinism entails the absurd contemporaneity of all events. And it simply begs the question to declare in this context that '*the passage of time*... is the very essence of the concept'.[23] For the undeniable fact that passage in the sense of transiency of the now is integral to the common-sense concept of time may show only that, in this respect, this concept is anthropocentric.

The becomingless physical world of the Minkowski representation is viewed *sub specie aeternitatis* in that representation in the sense that the relativistic account of time represented by it makes no reference to the particular times of anyone's *now*-perspectives. And, as J. J. C. Smart observed, 'the tenseless way of talking does not therefore imply that physical things or events are eternal in the way in which the number 7 is'.[24] We must therefore reject Whitrow's odd claim that according to the relativistic conception of Minkowski, 'external events *permanently* exist and we merely come across them'.[25] According to Minkowski's conception, an event qualifies as a *becomingless* occurrence by occurring in a network of relations of earlier and later and thus can be said to occur 'at a certain time *t*'. Hence to assert tenselessly that an event exists (occurs) is to claim that there is a time or clock reading *t* with which it coincides. But surely this assertion does not entail the absurdity that the event exists (occurs) at *all* clock times or 'permanently'. To occur tenselessly at some time *t* or other is not at all the same as to exist 'permanently'.

Whitrow himself acknowledges Minkowski's earlier-later relations when he says correctly that 'the relativistic picture of the world recognizes only a difference between earlier and later and not between past, present, and future'.[26] But he goes on to query: 'if no events *happen*, except our observations, we might well ask – why are the latter exceptional?'[27] I reply first of all: But Minkowski asserts that events happen tenselessly in the sense of occurring at certain clock times. And as for the exceptional status of the events which we register in observational awareness, I make the following obvious but only partial retort: being registered in awareness, these events are *eo ipso* exceptional.

I say that this retort is only partial because behind Whitrow's question there lurks a more fundamental query. This query must be answered by those of us who claim with Russell that 'past, present, and future arise from time-relations of subject and object, while earlier and later arise from time-relations of object and object'.[28] That query is: Whence the becoming in the case of mental events that become and are causally dependent on physical events, given that physical events themselves do not become independently of being perceived but occur tenselessly? More specifically, the question is: if our *experiences* of (extra and/or intradermal) physical events are causally dependent upon these events, how is it that the former *mental* events can properly qualify as being 'now', whereas the eliciting physical events *themselves* do not so qualify, and yet both kinds of events are (severally and collectively) alike related by quasi-serial relations of earlier and later?[29]

But, as I see it, this question does not point to refuting evidence against the mind-dependence of becoming. Instead, its force is to demand (a) the recognition that the complex mental states of judgmental awareness as such have distinctive features of their own, and (b) that the articulation of these features as part of a theoretical account of 'the place of mind in nature' acknowledges *what may be peculiar to the time of awareness.* That the existence of features peculiar to the time of awareness does not pose perplexities militating against the mind-dependence of becoming seems to me to emerge from the following three counter questions, which I now address to the critics:

(1) Why is the mind-dependence of becoming more perplexing than the mind-dependence of common-sense color attributes? That is, why is the former more puzzling than that physical events like the reflection of certain kinds of photons from a surface causally induce mental events like seeing blue which are qualitatively fundamentally different in some respects? In asking this question, I am *not* assuming that nowness is a *sensory quality* like red or sweet, but only that nowness and sensory qualities alike depend on awareness.

(2) Likewise assuming the causal dependence of mental on physical events, why is the mind-dependence of becoming more puzzling than the fact that the raw feel components of mental events, such as a particular event of seeing green, are not members of the *spatial* order of physical events?[30] Yet mental events and the raw feels ingredient in them are part of a time system of relations of earlier and later that comprises physical events as well.[31]

(3) Mental events must differ from physical ones in some respect qua being mental, as illustrated by their not being members of the same system of spatial order. Why then should it be puzzling that on the strength of the *distinctive* nature of conceptualized awareness and self-awareness, mental events differ further from physical ones with respect

to becoming, while both kinds of events sustain temporal relations of simultaneity and precedence?

What is the reasoning underlying the critics' belief that their question has the capability of pointing to the refutation of the mind-dependence of becoming? Their reasoning seems to me reminiscent of Descartes's misinvocation of the principle that there must be nothing more in the effect than is in the cause *à propos* of one of his arguments for the existence of God: the more perfect, he argued, cannot proceed from the less perfect as its efficient and total cause. The more perfect, i.e., temporal relations involving becoming, critics argue, cannot proceed from the less perfect, i.e., becomingless physical time, as its efficient cause. By contrast, I reason that nowness (and thereby pastness and futurity) are features of events *as experienced* conceptually, *not* because becoming is likewise a feature of the physical events which causally elicit our awareness of them, but because these elicited states are indeed specified states of *awareness*. Once we recognize the role of awareness here, then the diversity and order of the events of which we have awareness in the form of now-contents gives rise to the transiency of the now as explained in Section III above, due cautions being exercised, as I emphasized there, that this transiency not be construed tautologically.

In asserting the mind-dependence of becoming, I allow fully that the kind of neurophysiological brain state which underlies our mere awareness of an event as simply occurring now differs in specifiable ways from the ones underlying tick-tock or melody awareness, memory-awareness, anticipation-awareness, and dream-free sleep. But I cannot see why the states of awareness which make for becoming must have physical event-counterparts which isomorphically become in their own right. Hence I believe to have coped with Whitrow's question as to why only perceived events become. Indeed, it seems to me that the thesis of mind-dependence is altogether free from an important perplexity which besets the opposing claim that physical events are inherently past, present, and future. This perplexity was stated by Smart as follows: 'If past, present, and future were real properties of events [i.e., properties possessed by physical events independently of being perceived], then it would require [non-trivial] explanation that an event which becomes present [i.e., qualifies as occurring *now*] in 1965 becomes present [now] at that date and not at some other (and this would have to be an explanation over and above the explanation of why an event of this sort occurred in 1965)'.[32] It would, of course, be a complete trivialization of the thesis of the mind-*in*dependence of becoming to reply that *by definition* an event occurring at a certain clock time *t* has the unanalyzable attribute of nowness at time *t*.

Thus to the question 'Whence the becoming in the case of mental events that become and are causally dependent on physical events which do not themselves become?' I reply: 'Becoming can characterize mental events qua their being both bits of *awareness* and sustaining relations of temporal order'.

The awareness which each of several human percipients has of a given physical event can be such that all of them are alike prompted to give the same tensed description of the external event. Thus, suppose that the effects of a given physical event are simultaneously registered in the awareness of several percipients such that they each perceive the event as occurring at essentially the time of their first awareness of it. Then they may each think at that time that the event belongs to the present. The parity of access to events issuing in this sort of intersubjectivity of tense has prompted the common-sense belief that the nowness of a physical event is an intrinsic, albeit transient attribute of the event. But this kind of intersubjectivity does not discredit the mind-dependence of becoming; instead, it serves to show that the becoming present of an event, though mind-dependent no less than a pain, need not be *private* as a pain is. Some specific person's particular pain is private in the sense that this person has privileged access to its raw feel component.[33] The mind-dependence of becoming is no more refuted by such intersubjectivity as obtains in regard to tense than the mind-dependence of common-sense color attributes is in the least disproven by agreement among several percipients as to the color of some chair.

VI. BECOMING AND THE CONFLICT BETWEEN DETERMINISM AND INDETERMINISM

If the doctrine of mind-dependence of becoming is correct, a very important consequence follows, which seems to have been previously overlooked: Let us recall that the nowness of events is generated by (our) conceptualized *awareness* of them. Therefore, *nowness is made possible by processes sufficiently macro-deterministic (causal) to assure the requisitely high correlation between the occurrence of an event and someone's being made suitably aware of it.* Indeed, the very concept of experiencing an external event rests on such macro-determinism, and so does the possibility of empirical knowledge. In short, insofar as there is a transient present, it is made possible by the existence of the requisite degree of macro-determinism in the physical world. And clearly, therefore, the transiency of the present can obtain in a completely deterministic physical universe, be it relativistic or Newtonian.

The theory of relativity has repudiated the uniqueness of the simultaneity slices within the class of physical events which the Newtonian theory had affirmed. Hence Einstein's theory certainly precludes the conception of 'the present' which some defenders of the objectivity of becoming have linked to the Newtonian theory. But it must be pointed out that the doctrine of the mind-dependence of becoming, being entirely compatible with the Newtonian theory as well, does not depend for its validity on the espousal of Einstein's theory as against Newton's.

Our conclusion that there can be a transient now in a completely deterministic physical universe is altogether at variance with the contention of a number of distinguished thinkers that the indeterminacy of the laws of physics is both a necessary and sufficient condition for becoming. And therefore I now turn to the examination of their contention.

According to such noted writers as A. S. Eddington, Henri Bergson, Hans Reichenbach, H. Bondi, and G. J. Whitrow, it is a distinctive feature of an *indeterministic* universe, as contrasted with a deterministic one, that physical events belong to the present, occur *now*, or come into being over and above merely becoming present in *awareness*. I shall examine the argument given by Bondi, although he no longer defends it, as well as Reichenbach's argument. And I shall wish to show the following: insofar as events do become, the indeterminacy of physical laws is neither sufficient nor necessary for conferring nowness or presentness on the occurrences of events, an attribute whereby the events come into being. And thus my analysis of their arguments will uphold my previous conclusion that far from depending on the indeterminacy of the laws of physics, becoming requires a considerable degree of macro-determinism *and* can obtain in a completely deterministic world. Indeed, I shall go on to point out that not only the becoming of any kind of event but the temporal order of earlier and later among physical events depends on the at least quasi-deterministic character of the macrocosm. And it will then become apparent in what way the charge that a deterministic universe must be completely *timeless* rests on a serious misconstrual of determinism.

Reichenbach contends: 'When we speak about the progress of time [from earlier to later]..., we intend to make a synthetic [i.e., factual] assertion which refers both to an immediate experience and to physical reality'.[34] And he thinks that this assertion about events coming *into* being independently of mind – as distinct from merely occurring tenselessly at a certain clock time – can be justified in regard to physical reality on the basis of indeterministic quantum mechanics by the following argument:[35] In classical deterministic physics, both the past and the future were determined in relation to the present by one-to-one functions even though they differed in that there could be direct observational records of the past and only predictive inferences concerning the future. On the other hand, while the results of past measurements on a quantum mechanical system are *determined* in relation to the present *records* of these measurements, a present measurement of one of two conjugate quantities does *not* uniquely determine in any way the result of a *future* measurement of the other conjugate quantity. Hence, Reichenbach concludes:

> The concept of "becoming" acquires significance in physics: the present, which separates the future from the past, is the moment at which that which was undetermined becomes determined, and "becoming" has the same meaning as "becoming determined." ... it is with respect to "now" that the past is determined and that the future is not.[36]

I join Hugo Bergmann[37] in rejecting this argument for the following reasons. In the indeterministic quantum world, the relations between the sets of measurable values of the state variables characterizing a physical system at different times are, in principle, *not* the one-to-one relations linking the states of classically behaving closed systems. But I can assert correctly in 1966 that this holds for a given state of a physical system and its absolute future quite independently of whether that state occurs at midnight on December 31, 1800 or at noon on March 1, 1984. Indeed, if we consider *any one* of the temporally successive regions of space-time, we can veridically assert the following at *any* time: the events belonging to that particular region's absolute past could be (more or less) uniquely specified in records which are a part of that region, whereas its particular absolute future is thence quantum mechanically unpredictable. Accordingly, *every* event, be it that of Plato's birth or the birth of a person born in the year 2000 A.D., *at all times* constitutes a divide in Reichenbach's sense between its own recordable past and its unpredictable future, *thereby satisfying Reichenbach's definition of the 'present' or 'now' at any and all times!* And if Reichenbach were to reply that the indeterminacies of the events of the year of Plato's birth have already been transformed into a determinacy, whereas those of 2000 A.D. have not, then the rejoinder would be: this tensed conjunction holds for any state between sometime in 428 B.C. and 2000 A.D. that qualifies as now during that interval on grounds other than Reichenbach's asymmetry of determinedness; but the second conjunct of this conjunction does not hold for any state after 2000 A.D. which qualifies as now after that date. Accordingly, contrary to Reichenbach, the now of conceptualized awareness must be invoked tacitly at time *t*, if the instant *t* is to be nontrivially and nonarbitrarily singled out as present or now by Reichenbach's criterion, i.e., if the instant *t* is to be uniquely singled out at time *t* as being 'now' in virtue of being the threshold of the transition from indeterminacy to determinacy.

Turning to Bondi, we find him writing:

> ... the flow of time has no significance in the logically fixed pattern demanded by deterministic theory, time being a mere coordinate. In a theory with indeterminacy, however, the passage of time transforms statistical expectations into real events.[38]

If Bondi intended this statement to assert that the indeterminacy makes for our human inability to know in advance of their actual occurrence what particular kinds of events will in fact materialize, then, of course, there can be no objection. For in an indeterministic world, the attributes of specified kinds of events are indeed not uniquely fixed by the properties of earlier events and are therefore correspondingly unpredictable. But I take him to affirm beyond this the following traditional philosophical doctrine; in an indeterministic world, events come *into* being by becoming present with time, whereas in a deterministic world the status of events is one of merely occurring tenselessly at certain times. And my objections to his appeal to the transformation of statistical expectations into real events by the passage of time fall into several groups as follows.

(1) Let us ask: what is the character and import of the difference between a (micro-physically) indeterministic and a deterministic physical

world in regard to the attributes of future events? The difference concerns only the type of functional connection linking the attributes of future events to those of present or past events. Thus, *in relation to the states existing at other times*, an indeterministic universe allows alternatives as to the attributes of an event that occurs at some given time, whereas a deterministic universe provides no corresponding latitude. But this difference does *not* enable (micro-physical) indeterminism – as contrasted with determinism – to make for a difference in the *occurrence-status* of future events by enabling them to come *into* being. Hence in an indeterministic world, physical events no more *become* real (i.e., present) and are no more precipitated into existence, as it were, than in a deterministic one. In either a deterministic or indeterministic universe, events can be held to come into being or to become 'actual' by becoming *present in (our) awareness*; but becoming actual in virtue of occurring *now* in that way no more makes for a mind-independent coming into existence in an indeterministic world than it does in a deterministic one.

(2) Nor does indeterminacy as contrasted with determinacy make for any difference whatever at any time in regard to the *intrinsic attribute-specificity* of the future events themselves, i.e., to their being (tenselessly) what they are. For in either kind of universe, it is a fact of logic that what will be, will be, no less than what is present or past is indeed present or past![39] The result of a future quantum mechanical measurement may not be definite prior to its occurrence in relation to earlier states, and thus our prior knowledge of it correspondingly cannot be definite. But a quantum mechanical event has a tenseless occurrence status at a certain time which is fully compatible with its intrinsic attribute-definiteness just as a measurement made in a deterministic world does. Contrary to a widespread view, this statement holds also for those events which are constituted by energy states of quantum mechanical systems, since energy *can* be measured in an arbitrarily short time in that theory.[40]

Let me remark parenthetically that the quantum theory of measurement has been claimed to show that the *consciousness* of the human observer is essential to the definiteness of a quantum mechanical event. I have no firsthand knowledge of the technical details of the argument for this conclusion, but I hope that I shall be pardoned for nonetheless raising the following question in regard to it. Can the quantum theory account for the relevant physical events which presumably occurred on the surface of the earth *before* man and his consciousness had evolved? If so, then these physical events cannot depend on human consciousness for their specificity. On the other hand, if the quantum theory cannot in principle deal with *pre*-evolutionary physical events, then one wonders whether this fact does not impugn its adequacy in a fundamental way.

In an indeterministic world, there is a lack of attribute-specificity of events *in relation to events at other times*. But this *relational* lack of attribute-specificity cannot alter the fact of logic that an event is intrinsically attribute-specific in the sense of tenselessly being what it is at a certain clock time *t*.[41]

It is therefore a far-reaching mistake to suppose that unless and until an event of an indeterministic world belongs to the present or past, the event must be *intrinsically* attribute-*in*definite. This error is illustrated by Capek's statement that in the case of an event 'it is only its presentness [i.e., now-ness] which creates its specificity ... by eliminating all other possible features incompatible with it'.[42] Like Bondi, Capek overlooks that it is only with respect to some now or other that an event can be future at all to begin with and that the lack of attribute-specificity or 'ambiguity' of a future event is not intrinsic but relative to the events of the prior now-perspectives.[43] In an indeterministic world, an event is intrinsically attribute-determinate by being (tenselessly) what it is (tenselessly), regardless of whether the time of its occurrence be now (the present) or not. What makes for the coming into being of a future event at a later time *t* is *not* that its attributes are indeterministic with respect to prior times but only that it is registered in the now-content of awareness at the subsequent time *t*.

(3) Two quite different things also seem to be confused when it is inferred that in an indeterministic quantum world, future physical events themselves distinctively come into being with the passage of time over and above merely occurring and becoming present to awareness, whereas in a deterministic universe they do not come into being: (i) the epistemic precipitation of the *de facto* event-properties of future events out of the wider matrix of the possible properties allowed in advance by the quantum-mechanical probabilities, a precipitation or becoming definite which is constituted by our getting to *know* these *de facto* properties at the later times, and (ii) a mind-independent coming into being over and above merely occurring and becoming present to awareness at the later time.

The *epistemic* precipitation is indeed effected by the passage of time through the transformation of a merely statistical expectation into a definite piece of available information. But this does *not* show that in an indeterministic world there obtains any kind of becoming present ('real') with the passage of time that does not also obtain in a deterministic one. And in either kind of world, becoming as distinct from mere occurrence at a clock time requires conceptualized awareness.

We see then that the physical events of the indeterministic quantum world as such do not come into being anymore than those of the classical deterministic world but alike occur tenselessly. And my earlier contention that the transient now is mind-dependent and irrelevant to physical events as such therefore stands.

Proponents of indeterminism as a physical basis of objective becoming have charged that a deterministic world is timeless. Thus, Capek writes:

> ... the future in the deterministic framework ... becomes something *actually* existing, a sort of disguised and hidden present which remains hidden only from our limited knowledge, just as distant regions of space are hidden from our sight. "Future" is merely a label given by us to the unknown part of the *present* reality, which exists in the same degree as scenery hidden from our eyes. As this hidden portion of the present is *contemporary* with the portion accessible to us, the temporal relation between the present and the future is eliminated; the future loses its status of "futurity" because instead of succeeding the present it *coexists* with it.[44]

In the same vein, G. J. Whitrow declares:

> There is indeed a profound connection between the reality of time and the existence of an incalculable element in the universe. Strict causality would mean that the consequences pre-exist in the premisses. But, if the future history of the universe pre-exists logically in the present, why is it not already present? If, for the strict determinist, the future is merely "the hidden present," whence comes the illusion of temporal succession?[45]

But I submit that there is a clear and vast difference between the relation of one-to-one functional connection between two temporally-separated states, on the one hand, and the relation of temporal coexistence or simultaneity on the other. How, one must ask, does the fact that a future state is uniquely specified by a present state detract in the least from its being later and entail that it paradoxically exists at present? Is it not plain that Capek trades on an ambiguous use of the terms 'actually existing' and 'coexists' to confuse the time sequential relation of being *determined* by the present with the simultaneity relation of contemporaneity with the present? In this way, he fallaciously saddles determinism with entailing that future events exist now just because they are determined by the state which exists now. When he tells us that according to determinism's view of the future, 'we are already dead without realizing it now',[46] he makes fallacious use of the correct premiss that according to determinism, the present state uniquely specifies at what later time any one of us shall be dead. For he refers to the determinedness of our subsequent deaths misleadingly as our 'already' being dead and hence concludes that determinism entails the absurdity that we are dead *now*! Without this ambiguous construal of the term 'already', no absurdity is deducible.

When Whitrow asks us why, given determinism, the future is not already present even though it 'pre-exists logically in the present', the reply is: precisely because existing at the present time is radically different in the relevant temporal respect from what he calls 'logical pre-existence in the present'. Whitrow ignores the fact that states hardly need to be simultaneous just because they are related by one-to-one functions. And he is able to claim that determinism entails the illusoriness of temporal succession (i.e., of the earlier-later relations) only because he uses the term 'hidden present' just as ambiguously as Capek uses the term 'coexists'. But, more fundamentally, we have learned from the theory of relativity that events sustain *time-like* separations to one another *because* of their *causal* connectibility or deterministic relatedness, *not* despite that deterministic relatedness. And nothing in the relativistic account of the temporal order depends on the existence of an indeterministic microphysical substratum! Indeed, in the absence of the causality assumed in the theory in the form of causal (signal) connectibility, it is altogether unclear how the system of relations between events would possess the kind of *structure* that we call the 'time' of physics.[47]

VII. SUMMARY

In this lecture, I have presented my reasons for denying that nowness and temporal becoming are entitled to a place within physical theory, be it deterministic or indeterministic. On the other hand, the temporal relations of earlier than, later than, and of simultaneity do, of course, obtain among physical events in their own right in the sense familiar from the theory of relativity. Hence, if the 'meaning' of time is held to comprise becoming or passage, then one of the features of time is mind-dependent. But in characterizing becoming as mind-dependent, I allow fully that the mental events on which it depends themselves require a biochemical physical base or possibly a physical basis involving cybernetic hardware.

University of Pittsburgh

REFERENCES

* A Louis Clark Vanuxem Lecture, delivered at Princeton University on March 2, 1967. This paper includes a *revised* version of Chapter I of my *Modern Science and Zeno's Paradoxes*, Wesleyan University Press, Middletown, 1967; 2nd ed., Allen & Unwin Ltd., London, 1968.

1 G. J. Whitrow, *The Natural Philosophy of Time*, Thomas Nelson & Sons, Ltd., London, 1961, p. 88.

2 For example, consider the events in the careers of human beings or of animals who *return* to a spatially fixed terrestrial habitat every so often. These events occur at space points on the earth which certainly do *not* exhibit the betweenness of the points on a Euclidean straight line.

3 For details, cf. Adolf Grünbaum, *Philosophical Problems of Space and Time*, Alfred A. Knopf, Inc., New York, 1964, pp. 214–216. Hereafter this work will be cited as *PPST*.

4 This noncommittal character of the term 'initial state' seems to have been recognized by O. Costa de Beauregard in one part of his paper entitled 'Irreversibility Problems', *Logic, Methodology and Philosophy of Science*, Proceedings of the 1964 International Congress (ed. by Y. Bar-Hillel), North-Holland Publ. Co., Amsterdam, 1965, p. 327. But when discussing my criticism of Hans Reichenbach's account of irreversibility (*PPST*, pp. 261–263), Costa de Beauregard (*ibid.*, p. 331) overlooks that my criticism invokes initial states in only the noncommittal sense set forth above.

5 Thus, it is erroneous to maintain, as Milic Capek does, that the distinction between temporal betweenness and irreversibility is 'fallacious' in virtue of being 'based on the superficial and deceptive analogy of "the course of time" with a geometrical line', *The Philosophical Impact of Contemporary Physics*, D. Van Nostrand Co., Inc., Princeton, New Jersey, 1961, p. 349; see also pp. 347 and 355. If Capek's condemnation of this distinction were correct, the following fundamental question of theoretical physics could not even be intelligibly and legitimately asked: Are the *prima facie* irreversible processes known to us indeed irreversible, and, if so, on the strength of what laws and/or boundary conditions are they so? For this question is predicated on the very distinction which Capek rejects as 'fallacious'. By the same token, Capek errs (*ibid.*, p. 355) in saying that when Reichenbach characterizes entropically counterdirected epochs as 'succeeding each other', then irreversibility 'creeps in' along with the asymmetrical relations of before and after. For all that he needs to assume here to speak of 'before' and 'after' is a time coordinatization which reflects the assumed kind of betweenness and simultaneity.

6 On the basis of a highly equivocal use of the term 'irreversible', M. Capek, *ibid.*, pp. 166–167 and 344–345 has claimed incorrectly that the account of the space-time properties of world lines given by the special theory of relativity entails the irreversi-

bility of physical processes represented by world lines. He writes: 'The world lines, which by definition are constituted by a succession of isotopic events, are *irreversible* in all systems of reference' (*ibid.*, p. 167) and 'the relativistic universe is dynamically constituted by the network of causal lines *each of which is irreversible*; ... this irreversibility is a topological invariant' (*ibid.*, pp. 344–345). But Capek fails to distinguish between (1) the *non-inversion or invariance of time-order as between different Galilean frames* which the Lorentz-transformation equations assert in the case of causally connectible events, and (2) the irreversibility of processes represented by world lines in the standard sense of the *non-restorability* of the same kind of state in any frame. Having applied the term 'irreversibility' to (1) no less than to (2) after failing to distinguish them, Capek feels entitled to infer that the Lorentz transformations attribute irreversibility within any one frame to processes depicted by world lines, just because these transformations assert the invariance of time order on the world lines as between different frames. That the Lorentz equations do not disallow the reversibility of physical processes becomes clear upon making each of the *two* replacements $t \to -t$ and $t' \to -t'$ in them: these replacements issue in the same set of equations except for the sign of the velocity term in each of the numerators, i.e., they merely reverse the direction of the motion. Therefore, these two replacements do *not* involve any violation of the theory's time-order invariance as between different frames S and S'. By contrast, different equations exhibiting a violation of time-order invariance on the world lines would be obtained by replacing *only* one of the two variables t and t' by its negative counterpart in the Lorentz equations.

[7] For a discussion of the various kinds of irreversible processes which make for the anisotropy of time and furnish specified criteria for the relations of temporal precedence and succession, see Costa de Beauregard, *op. cit.*, p. 327; and A. Grünbaum, *PPST*, Ch. 8, and 'The Anisotropy of Time', in *The Nature of Time* (ed. by D. L. Schumacher and T. Gold), Cornell University Press, Ithaca, 1967, pp. 149–186.

[8] Some have questioned the possibility of stating what specific physical events do occur in point of fact at particular clock times without covert appeal to the transient now. Cf. Hermann Weyl, *Philosophy of Mathematics and Natural Science*, Princeton University Press, Princeton, 1949, p. 75. In their view, any physical description will employ a time coordinatization, and any such coordinatization must ostensively invoke the now to designate at least one state as, say, the origin of the time coordinates. But I do not see a genuine difficulty here for three reasons. Firstly, it is not clear that the designation of the birth of Jesus, for example, as the origin of time coordinates tacitly makes logically indispensable use of the now or of tenses in virtue of making use of a proper name. Secondly, in some cosmological models of the universe, an origin of time coordinates can clearly be designated non-ostensively: in the 'big bang' model, the big bang itself can be designated uniquely and *non*-ostensively as the one state having no temporal predecessor. And thirdly, any two descriptions of the world which differ only in the choice of the origin of time coordinates while employing the same time metric and time topology are equivalent with respect to their factual *physical* content. Thus such descriptions differ only in regard to the way in which they numerically name or label particular simultaneity classes of events. Hence, let us grant for argument's sake that tacit use of the now or of tenses is logically indispensable to designating the origin of any one particular time coordinatization. Even if this is granted, it does *not* follow that past, present, and future have a mind-independent status in the temporal structure of the physical world.

[9] The claim that the now advances in the direction of the future is a truism as regards

both the correspondence between nows and *physically* later clock times and their correspondence with psychologically (introspectively) later contents of awareness. What is *not* a truism, however, is that the *introspectively* later nows are *temporally correlated* with states of our physical environment that are later as per criteria furnished by irreversible physical processes. This latter correlation depends for its obtaining on the laws governing the physical and neural processes necessary for the *mental* accumulation of memories and for the registry of information *in awareness*. (For an account of some of the relevant laws, see A. Grünbaum, *PPST*, Ch. 9, Secs. A and B.) Having exhibited the aforementioned truisms as such and having noted the role of the empirical laws just mentioned, I believe to have answered Costa de Beauregard's complaint (in 'Irreversibility Problems', *op. cit.*, p. 337) that 'stressing that the arrows of entropy and information increase are parallel to each other is *not* proving that the flow of subjectivistic time has to follow the arrows!'

[10] A very illuminating account of the logical relations of Minkowski's language to tensed discourse is given by Wilfrid Sellars in 'Time and the World Order', *Minnesota Studies in the Philosophy of Science*, vol. III (ed. by H. Feigl and G. Maxwell), University of Minnesota Press, Minneapolis, 1962, p. 571.

[11] Quoted from G. J. Whitrow, *op. cit.*, pp. 129–130.

[12] For a searching treatment of the ramifications of the contrast pertinent here, see Wilfrid Sellars, 'Philosophy and the Scientific Image of Man', *Frontiers of Science and Philosophy* (ed. by Robert G. Colodny), University of Pittsburgh Press, Pittsburgh, 1962, pp. 35–78.

[13] It will be noted that I speak here of the dependence of nowness on an organism *M* which is mind-possessing in the sense of having conceptualized or judgmental awareness, as contrasted with mere sentiency. Since biological organisms other than man (e.g., extra-terrestrial ones) may be mind-possessing in this sense, it would be unwarrantedly restrictive to speak of the mind-dependence of nowness as its 'anthropocentricity'. Indeed, it might be that conceptualized awareness turns out not to require a *biochemical* substratum but can also inhere in a suitably complex 'hardware' computer. That a physical substratum of some kind is required would seem to be abundantly supported by the known dependence of the content and very existence of consciousness in man on the adequate functioning of the human body.

[14] The distinction pertinent here between the *mere* hearing of something and judgmental awareness that it is being heard is well stated by Roderick Chisholm as follows: 'We may say of a man simply that he observes a cat on the roof. Or we may say of him that he observes *that* a cat is on the roof. In the second case, the verb "observe" takes a "that"-clause, a propositional clause as its grammatical object. We may distinguish, therefore, between a "propositional" and a "nonpropositional" use of the term "observe", and we may make an analogous distinction for "perceive", "see", "hear", and "feel"'.

'If we take the verb "observe" propositionally, saying of the man that he observes that a cat is on the roof, or that he observes a cat to be on the roof, then we may also say of him that he *knows* that a cat is on the roof; for in the propositional sense of "observe", observation may be said to imply knowledge. But if we take the verb nonpropositionally, saying of the man only that he observes a cat which is on the roof, then what we say will not imply that he knows that there is a cat on the roof. For a man may be said to observe a cat, to see a cat, or hear a cat, in the nonpropositional sense of these terms, without his knowing that a cat is what he is observing, or seeing, or hearing. "It was not until the following day that I found out that what I saw was only a cat"'. *Theory of Knowledge*, Prentice-Hall, Inc., Englewood Cliffs, New Jersey, 1966, p. 10. I am indebted to Richard Gale for this reference.

[15] The judgmental awareness which I claim to be essential to an event's qualifying as now may, of course, be expressed by a linguistic utterance, but it clearly need not be so expressed. I therefore consider an account of nowness which is *confined* to utterances as inadequate. Such an overly restrictive account is given in J. J. C. Smart's otherwise illuminating defense of the anthropocentricity of tense, *Philosophy and Scientific Realism*, Routledge & Kegan Paul, London, 1963, Chapter vii. But this undue restrictiveness is quite inessential to his thesis of the anthropocentricity of nowness. And the non-restrictive treatment which I am advocating in its stead would obviate his having to rest his case on (1) denying that 'this utterance' can be analyzed as 'the utterance which is *now*', and (2) insisting that 'now' must be elucidated in terms of 'this utterance' (*ibid.*, pp. 139–140).

[16] Hans Reichenbach, *The Direction of Time*, University of California Press, Berkeley, Calif., 1956, p. 16.

[17] Paul Fraisse, *The Psychology of Time*, Eyre & Spottiswoode, London, 1964, p. 73.

[18] A. Grünbaum, *PPST*, p. 325.

[19] M. Capek, *op. cit.*, p. 337.

[20] On the basis of such a misunderstanding, M. Capek incorrectly charges the thesis with a 'spatialization of time' in which 'successive moments already *coexist*' (*ibid.*, pp. 160–163) and in which 'the universe with its whole history is conceived as a single huge and timeless bloc, given at once' (*ibid.*, p. 163). See also p. 355.

[21] G. J. Whitrow, *op. cit.*, p. 228 (my italics). For a criticism of another such misconstrual, see A. Grünbaum, *PPST*, pp. 327–328.

[22] G. J. Whitrow, *ibid.*, p. 88.

[23] *Ibid.*, pp. 227–228.

[24] J. J. C. Smart, *op. cit.*, p. 139.

[25] G. J. Whitrow, *op. cit.*, p. 88, n. 2 (my italics).

[26] *Ibid.*, p. 293.

[27] *Ibid.*, p. 88, n. 2.

[28] Bertrand Russell, 'On the Experience of Time', *The Monist* **25** (1915), 212.

[29] The need to deal with this question has been pointed out independently by Donald C. Williams and Richard Gale.

[30] Mental events, as distinct from the neurophysiological counterpart states which they require for their occurrence, are *not* in our heads in the way in which, say, a biochemical event in the cortex or medulla oblongata is.

[31] Thus a conscious state of elation induced in me by the receipt of good news from a telephone call C_1 could be *temporally between* the physical chain C_1 and another such chain C_2 consisting of my telephonic transmission of the good news to someone else.

[32] J. J. C. Smart, *op. cit.*, p. 135.

[33] I am indebted to Richard Gale for pointing out to me that since the term 'psychological' is usefully reserved for mind-dependent attributes which are private, as specified, it would be quite misleading to assert the mind-dependence of tense by saying that tense is 'psychological'. In order to allow for the required kind of intersubjectivity, I have therefore simply used the term 'mind-dependent'.

[34] Hans Reichenbach, *The Philosophy of Space and Time*, Dover Publications, Inc., New York, 1958, pp. 138–139.

[35] Hans Reichenbach, 'Les Fondements Logiques de la Mécanique des Quanta', *Annales de l'Institut Poincaré* **13** (1953), 154–157.

[36] *Ibid.*
[37] Cf. H. Bergmann, *Der Kampf um das Kausalgesetz in der jüngsten Physik*, Vieweg & Sohn, Braunschweig, 1929, pp. 27–28.
[38] H. Bondi, 'Relativity and Indeterminacy', *Nature* **169** (1952), 660.
[39] I am indebted to Professor Wilfrid Sellars for having made clarifying remarks to me in 1956 which relate to this point. And Costa de Beauregard has reminded me of the pertinent French dictum *Ce qui sera, sera.*
[40] Yakir Aharonov and David Bohm have noted that time does not appear in Schrödinger's equation as an operator but only as a parameter and have pointed out the following: (1) The time of an energy state is a dynamical variable belonging to the measuring apparatus and therefore *commutes* with the energy of the observed system. (2) Hence the energy state and the time at which it exists do *not* reciprocally limit each other's well-defined status in the manner of the non-commuting conjugate quantities of the Heisenberg Uncertainty Relations. (3) Analysis of illustrations of energy measurement (e.g., by collision) which seemed to indicate the contrary shows that the experimental arrangements involved in these examples did not exhaust the measuring possibilities countenanced by the theory. Cf. their two papers on 'Time in the Quantum Theory and the Uncertainty Relation for Time and Energy', *Physical Review* **122** (1961), 1649, and *Physical Review* **134** (1964), B1417. I am indebted to Professor A. Janis for this reference.
[41] A helpful account of the difference relevant here between being *determinate* (i.e., intrinsically attribute-specific) and being *determined* (in the relational sense of causally necessitated or informationally ascertained), is given by Donald C. Williams in *Principles of Empirical Realism*, Charles C. Thomas, Springfield, Ill., 1966, pp. 274 ff.
[42] Capek, *op. cit.*, p. 340.
[43] Capek writes further: 'As long as the ambiguity of the future is a mere appearance due to the limitation of our knowledge, the temporal character of the world remains necessarily illusory', and 'the principle of indeterminacy ... means the *reinstatement of becoming in the physical world*' [*ibid.*, p. 334]. But granted that the indeterminacy of quantum theory is ontological rather than merely epistemological, this indeterminacy is nonetheless relational and hence unavailing as a basis for Capek's conclusions.
[44] *Ibid.*, pp. 334–335, cf. also p. 164.
[45] G. J. Whitrow, *op. cit.*, p. 295.
[46] M. Capek, *op. cit.*, p. 165.
[47] Accordingly, we must qualify the following statement by J. J. C. Smart, *op. cit.*, pp. 141–142: 'We can now see also that the view of the world as a space-time manifold no more implies determinism than it does the fatalistic view that the future "is already laid up". It is compatible both with determinism and with indeterminism, i.e., both with the view that earlier time slices of the universe are determinately related by laws of nature to later time slices and with the view that they are not so related'. This statement needs to be qualified importantly, since it would not hold if 'indeterminism' here meant a macro-indeterminism such that macroscopic causal chains would not exist.

NICHOLAS RESCHER

LAWFULNESS AS MIND-DEPENDENT

I. WHAT IS A UNIVERSAL LAW? THE NATURE OF LAWFULNESS

Scientific explanations are subsumption arguments: they place special cases within a framework of regularity represented by laws. Our concept of explanation – causal explanation preeminently included – is such as to require that the generalizations used for explanatory purposes must be *lawful*.

It is quite clear that not just any universal empirical generalization will qualify as a law in this scientific context of discussion – no matter how well established it may be. It is critically important to distinguish here between *accidental* generalizations on the one hand and *lawful* generalizations on the other. 'All coins in my pocket weigh less than one ounce' and 'All American presidents are natives of the continental United States' are examples of accidental generalizations. By contrast, generalizations like 'All elm trees are deciduous', 'All (pure) water freezes at 32°F' and 'All *y* chromosomes self-duplicate under stimulation' are lawful. An accidental generalization claims merely that something *is* so – perhaps even that it is always so, whereas a lawful generalization claims that something *must* (in some appropriate sense) be so.

Thus consider the following two answers to the explanatory question: 'Why did that tree shed its leaves last fall?'

(1) Because it is an elm, and *all elms are deciduous.*

(2) Because it is a tree in Smith's yard, and *all trees in Smith's yard are deciduous.*

The drastic difference in the satisfactoriness of these two 'explanations' is due exactly to the fact that the generalization deployed in the first is lawful, whereas that in the second is not.

Laws are akin to, yet different from both rules and descriptions. Like rules, laws state how things 'must be', yet unlike most familiar rules laws admit no exceptions, but are always 'obeyed'. Like descriptions, laws state how things are; yet unlike standard descriptions, laws go beyond describing how things in fact are to make claims about how they must be. Thus laws have both a descriptive and a rulish aspect that prevents their being grouped squarely into either category.

But just what is this factor of lawfulness that is present with generalizations and absent with others? The best way to answer this question of what lawfulness *is* is by inquiring into what it *does*. Lawfulness manifests itself in two related ways: *nomic necessity* and *hypothetical force*. Nomic necessity introduces the element of *must*, of inevitability. In asserting it *as a law* that 'All *A*'s are *B*'s' ('All timber wolves are carnivorous') we claim that the world being as it is, it is necessary that an *A* must be a *B* (i.e., that a timber wolf will under appropriate circumstances unfailingly develop as a meat-eating animal).

This nomic necessity manifests itself most strikingly in the context of hypothetical suppositions – especially counterfactual hypotheses. In accepting 'All *A*'s are *B*'s' ('All spiders are eight-legged') as a law, we have to be prepared to accept the conditional 'If *x* were an *A*, then *x* would be a *B*'. (If this beetle were a spider [which it is not], then it would have eight legs.) It is preeminently this element of hypothetical force that distinguishes a genuinely lawful generalization from an accidental generalization like 'All coins in my pocket weigh less than one half ounce'. For we would not be prepared to accept the conditional 'If this Venetian florin were a coin in my pocket, then it would weigh less than one half ounce'.[1]

The fact is that the statement

(1) All *X*'s are *Y*'s

makes a stronger claim when put forward as a law than when put forward as a 'mere' generalization. For if (1) is construed as a law it asserts 'All *X*'s *have to be* *Y*'s', that is, we obtain the stronger *nomological* generalization:

(1a) All *X*'s are *Y*'s *and further* if *z* (which is not an *X*) were an *X*, then *z* would be a *Y*.[2]

When a generalization is taken as lawful it obtains added force: it gains a further assertive increment – even though this nomic necessity will express itself primarily in applications of a counterfactual kind. For it is clearly in hypothetical and counterfactual contexts that nomic necessity manifests itself most strikingly.

Consider the counterfactual supposition: *Assume that this wire (which is actually made of copper) were made of rubber*. This supposition occurs in the following context:

Items of Knowledge

Facts: (1) This wire is made of copper.
(2) This wire is not made of rubber.
(3) This wire conducts electricity.

Laws: (4) Copper conducts electricity.
(5) Rubber does not conduct electricity.

Hypothesis: Not-(2), i.e., This wire is made of rubber.

To restore consistency in our knowledge in the face of this hypothesis we must obviously give up (1) and (2). But this is not sufficient. One of (5) or (3) must also go, so that *prima facie* we could adopt either of the conditionals:

(A) If this (copper) wire were made of rubber then it would not conduct electricity (because rubber does not conduct electricity).

(B) If this (copper) wire were made of rubber, then rubber would conduct electricity (because this wire conducts electricity).

That is, we get a choice between retaining (5) with alternative (A) and retaining (3) with alternative (B). It is precisely because classing a statement as a 'law' represents an epistemic commitment to retain it in the face of counterfactual hypotheses that the conditional (A), viz., 'If this wire were made of rubber, then it would not conduct electricity' strikes us as natural *vis-à-vis* (B).[3]

One effective way to motivate the distinction between a 'law' and a 'mere generalization' is to consider the effect of contraposition. The fact is that the statement

(1) All X's are Y's

makes a stronger claim when put forward as a law than when put forward as a 'mere' generalization. For if (1) is construed as a law it asserts 'All X's *have to be* Y's', that is, we obtain the stronger *nomological* generalization:

(1a) All X's are Y's *and further* if z (which is not an X) were an X, then z would be a Y.

When a generalization of the type (1) is taken in this nomological way, contraposition clearly fails. For

(2) All non-Y's are non-X's

when construed as stating a law, will assert:

(2a) All non-Y's are non-X's *and further* if z (which is not a non-Y) were a non-Y, then z would be a non-X.

Although the generalizations (1) and (2) are equivalent, this is not the case with their nomological counterparts (1a) and (2a). (These statements are nonequivalent because, *inter alia*, (1a) affirms that the z at issue in it is to be a non-X, while (2a) affirms that the z at issue in it is to be a Y [and neither statement justifies relating non-X's and Y's].) Thus *seemingly equivalent generalizations can formulate different, nonequivalent laws*. When a generalization is taken as nomological, that is as stating a law, it obtains an assertive increment (albeit one of a strictly counterfactual sort) of such a kind that contraposed generalizations will no longer represent the same law.

It is worth noting, incidentally, that these considerations help also to shed some light upon the inductive procedures by which laws are verified. For the purposes of inductive confirmation (1a) comes to be reconstrued in strictly factualistic, albeit epistemologized terms. It is reoriented from the realm of *counterfact* to that of *ignorance* (absence of information), being rendered epistemological rather than counterfactual by means of the indicated italicized insertions:

(1b) All X's are Y's and further if z (which is not *known to be* an X) were *to turn out* to be an X, then z would *turn out to* be a Y.

Thus in verifying (1) the *prima facie* procedure is to hunt for X's and check that they are Y's. On the other hand (2) gives rise to:

(2b) All non-Y's are non-X's and if z (which is not known to be a non-Y) were to turn out to be a non-Y, then z would turn out to be a non-X.

And so in verifying (2), the *prima facie* procedure is to hunt for non-Y's and check that they are non-X's. And these procedures, *qua* procedures, are very different, *despite the fact that a counterexample provided by one of them must also turn out to be a counterexample for the other* (viz., an X that is a non-Y). For although these generalizations are *refuted* (counter-instanced) by the same negative findings, they differ in the amount of the inductive *support* (confirmation) which they derive from positive, non-refuting instances of different kinds, and thus differ with respect to the pragmatics of verification.

It is built into our very concept of a law of nature that such a law must – if it be of the universal type – correspond to a universal generalization that is claimed to possess nomic necessity and is denied to be of a possible merely accidental status. If the generalization were claimed to hold *in fact* for all places and times, even this would not of itself suffice for lawfulness: it would still not be a law if its operative effectiveness were not also extended into the hypothetical sphere. The conception of a universal law operative in our concept of causal explanation is thus very complex and demanding. A lawful generalization goes beyond the claims of a merely factual generalization as such: it involves claims not only about the realm of observed fact, but about that of hypothetical counterfact as well. And just these far-reaching claims are indispensable to the acceptance of a generalization as lawful and represent a formative constituent of our standard concept of a universal law of nature.

That laws demand nomic necessity is a point regarding which there is a substantial concensus in the history of philosophy. Aristotle insists on the matter in the *Posterior Analytics*.[4] It is a basic theme in Kant's *Critique of Pure Reason*.[5] And it continues operative in current writers. Such authors as C. J. Ducasse and A. Pap, for example, hold that natural laws involve a necessity that is not logical but yet transcends merely *de facto* regularity.[6] Both Nelson Goodman and Roderick M. Chisholm have proposed hypothetical force as a prime criteria of the nomic necessity requisite for lawfulness.[7] And nowadays it is a matter of widespread agreement that some characteristic mode of nomic necessity is involved in lawfulness, although writers differ as to just how the factor of nomic necessity is to be explicated. The writer of the relevant article in the most recent philosophical encyclopedia puts the matter accurately by saying that the current point of dispute 'is not about the property of using such terms as "nomic necessity", rather it is about the interpretation of these terms or the justification of their use'.[8]

Conceding hypothetical force as an ingredient in all *universal* laws of nature, one recent writer denies its applicability to lawfulness in general on the grounds that it is lacking in statistical laws.[9] Now if I use that statistical law 'The halflife of californium is 5 hours' as basis for explaining why a particular atom of californium lasted $5\frac{1}{2}$ hours, then I cannot go on to say things like 'If this atom of uranium had been californium, then it would have lasted only $5\frac{1}{2}$ hours'. This sort of specific and nonprobabilistic application of a statistical law is indeed impossible. But we surely can and would be prepared to say things like 'If this atom of uranium had been californium, then it would have had a halflife of 5 hours, and so the probability is .85 that it would have lasted only 5 ± 1 hours'. Statistical laws too can (and indeed *must*, if lawful) be capable of counterfactual applications – only in their case such applications will, naturally enough, take a probabilistic form.

Some recent writers have opted for a 'regularity theory' of laws according to which lawfulness is to be construed as unrestricted factual generality pure and simple, so that no trans-inductive imputation of nomic necessity is called for. As R. S. Walters puts the matter, the key

> objection to the regularity theory is that it cannot account for possible instances. If this charge were indeed well founded, it would be difficult to see how one could avoid the view that natural laws assert some kind of necessity such that they apply in all possible worlds. However, it is not established that a defender of the regularity view cannot give a plausible account of the application of laws to possible instances. He would argue that statements about possible instances stand in the same kind of logical relation to a law as do statements about actual unobserved instances. To the extent that a law enables prediction about unobserved instances, it enables justifiable claims about unrealized possibilities.[10]

But it is quite clear that this line of defense will not serve at all. It suffers from the critical defect of treating the unobserved and the unactualized cases in exactly the same way. But it is quite clear on the basis of considerations we have already canvassed that this step is indefensible because the unobserved and the unreal are in a totally different position in the context of inductive considerations, since the realm of the (heretofore) unobserved lies open to observational exploration whereas the domain of the hypothetically unreal lies *ex hypothei* beyond our reach.

Of course one could try to argue that the consideration of hypothetical cases is improper or illicit (illegitimate, 'beside the point', or whatever) – that reality alone concerns us and that the unreal lies wholly outside the sphere of legitimate consideration. This does indeed abrogate the difficult nomic aspect of lawfulness. But it also writes off the prospect of hypothetical reasoning in the sciences and abolishes the concept of explanation as it has in fact developed in the context of the Western tradition of scientific methodology.[11]

II. LAWFULNESS AS IMPUTATION

On what evidential basis does an empirical generalization acquire nomic necessity and hypothetical force? However substantial this evidential basis may be – no matter how massively the observational evidence may be amassed and how elaborately the case may be developed – it is clear upon reflection that this evidential basis must always be grossly insufficient to the claim made when we class a generalization as a law. This becomes evident in part for the familiar reasons that while all such evidence relates to the past and possibly the present, scientific laws invariably also underwrite claims about the future. It is, moreover, also clear from considering the conceptual nature of lawfulness, bearing in mind that observation and evidence always relate to what happens in fact, whereas laws invariably also underwrite claims of a hypothetical or counterfactual kind.

Let us consider this matter of the sufficiency of evidential basis for a law somewhat more closely. It is obvious that this basis will be *deductively insufficient* because the evidence inevitably relates to a limited group of cases while the applicability of the law is unrestricted. Moreover the evidential basis will also be *inductively insufficient*. For inductive procedures are designed to warrant the step from observed to unobserved cases, whereas a law – whose very lawfulness arrogates to it nomological necessity and counterfactual force – takes not only this inductive step from observed to unobserved cases, but also takes the added step from actual to hypothetical cases. The inductive justification of hypothetical force would have to take the form 'has always been applicable to counterfactual cases' to 'will always be applicable to counterfactual cases'. And the premiss for such an induction will obviously always be unavailable. The evidential foundation for generalization is thus afflicted by a double insufficiency, not only in the *deductive* mode, but also *inductively* (at any rate as long as induction is construed along usual and standard lines).[12]

The basic fact of the matter – and it is a fact whose importance cannot be overemphasized – is that the elements of nomic necessity and hypothetical force are not to be extracted from the evidence: they are not *discovered* on some basis of observed fact at all – they are *supplied.* The realm of hypothetical counterfact is inaccessible to observational or experimental exploration.[13] Lawfulness is not found in or extracted from the evidence, it is superadded to it. *Lawfulness is a matter of imputation*: when an empirical generalization is designated as a law, this epistemological status is *imputed* to it. Lawfulness is something which a generalization could not *in principle* earn on the basis of warrant by the empirical facts. Men impute lawfulness to certain generalizations by according to them a particular role in the epistemological scheme of things, being prepared to use them in special ways in inferential contexts (particularly hypothetical contexts), and the like.

When one looks at the explicit formulation of the overt *content* of a law all one finds is a certain generalization. Its lawfulness is not a part of what the law asserts at all – it is nowhere to be seen in its overtly expressed content as a generalization. Lawfulness is not a matter of what the generalizations *says*, but a matter of *how it is to be used.* By being prepared to put it to certain kinds of uses in modal and hypothetical contexts, it is we the users who accord to a generalization its lawful status thus endowing it with nomological necessity and hypothetical force. Lawfulness is thus not a matter of the assertive content of a generalization, but of its epistemic status, as determined by the ways in which it is deployed in its applications.

This approach to lawfulness as imputed rests on a concept of the nature of scientific laws to which more explicit articulation must be given. Present-day philosophers of science have concentrated their attention primarily upon two aspects of 'laws': (1) their *assertive characteristics*, having to do with the machinery deployed in their formulation (they must be universal generalizations, must make no explicit reference to time, must contain no overt spatial delimitation, they should be 'simpler' than equally eligible alternatives, etc.), and (2) their *evidential status*, having to do with their supporting data (they must have no known counterinstances, should be supported by an ample body of confirming evidence, etc.).[14] To considerations of this sort one may add yet a third factor which could be put under the heading of an appropriate *epistemic commitment*, having to do with the extent to which we are committed to retention of the law in the face of putative discordant considerations of a strictly hypothetical character (and thus not of an evidential sort, for this would lead back to item (2) above) leading to a choice between it and other items of 'our knowledge'. The appropriateness of such epistemic commitment revolves about questions of the type: 'To what extent is the "law" at issue justifiably regarded as immune to rejection in the face of hypothetical considerations?' 'How should this generalization fare if (*per improbabile*) a choice were forced upon us between it and other laws we also accept?' 'How critical is it that the law be true – how serious a matter would it be were the law to prove false?'

This third factor represents an aspect of laws *crucially important to their status as laws*. For no matter what the structure of a generalization might be, or how well established it is by the known data, its acceptance as a law demands some accommodation of it within the 'system' of knowledge. Any 'law' occupies a place that is more or less fundamental within the general architectonic of our knowledge about the world – its epistemic status is a matter not only of *its own* form and *its own* evidential support, but of *its placement within the woof and warp of the fabric comprising it together with other cognate laws of nature*. The standing accorded to it within the overall framework of our knowledge reflects our 'epistemic commitment' to the law, which is thus a matter not of the individual characteristics of the 'law' viewed (insofar as possible) in isolation, but of its interconnections with and its relative epistemic entrenchment among other laws to which we are also committed. We must *decide* upon the epistemic status or ranking of the law with respect to others, and this decision, while in part guided by evidential considerations, is by no means determined by them alone, but is a matter of the entire range of systematic and systematic considerations, among which evidential considerations are only one (though to be sure a prominent) factor.

While it is, of course, 'we' who 'decide' upon the placement of a law in the epistemological scheme of things, and 'we' who make an 'epistemic commitment' to the law, the crucial point is that this be done on the basis of rational grounds (of complex and varied character) and not on the basis of a merely random choice or personal predilection. The appropriateness of epistemic commitment to laws is therefore not a matter of psychology or of the sociology of scientific knowledge. (It is necessary and proper to distinguish between being *in fact* committed to accepting a generalization as a 'law' of more or less fundamental status upon the one hand, and being *properly* or *warrantedly* committed to it, upon the other.)

In saying that the necessity and hypotheticality of lawfulness are matters of imputation, we have no intention whatsoever of suggesting that the issue is one of different conventions or arbitrary decisions. The imputation is, to be sure, an overt rational step for which a decision is required. But to be justified this decision must be based upon a rational warrant. It must have a grounding in (1) the *empirical evidence* for the generalization at issue in the law and (2) the *theoretical context* of the generalization. Such grounding is required to provide the necessary *warrant* to justify an imputation of lawfulness. Since an element of imputation is involved, laws are not just discovered, they are, strictly speaking, made. This is not, of course, to say they are made arbitrarily. Although they cannot be extracted from the empirical evidence, they must never violate it. Such conformity with 'the observed facts' is a key factor of that complex that bears the rubric of *well-foundedness*. Our conception of the origin of the key requisites for a low (nomic necessity and fact-transcending hypotheticality) can thus be summarized in the slogan: *Lawfulness is the product of the well-founded imputation to empirical generalizations of nomic necessity and hypothetical force.*

We must pursue somewhat further this key theme of the warrant for imputations of lawfulness, which we have held to be a question of evidence and of theoretical context. The factor of the evidence for generalizations may at this time of day be supposed to be relatively familiar to the reader. Even elementary discussions of scientific method devote considerable attention to the issue of the evidence needed for scientific laws. On the other hand, the bearing of the theoretical context of an empirical generalization in establishing its claims to lawfulness is a perhaps much less familiar issue.

It is not for nothing that branches of science are called *bodies* of knowledge. Scientific knowledge has a complex and highly articulated structure. The laws comprising this structure rarely if ever stand isolated and alone: they are part of a fabric whose threads run off to intertwine with other laws. Scientific laws do not stand in splendid isolation, they interlink with others in the complex logical structures commonly called *theories*. To say this is not to deny that there can be such things as 'merely empirical generalizations' – universal propositions which, though well-confirmed by the empirical evidence, wholly lack a footing within some ramified theoretical framework. Kepler's laws of planetary motion, Galileo's law of falling bodies, and Boyle's gas law, for example, were all well-established and generally accepted prior to securing the grounding provided by a foundation upon some theory. But an aggregation of well-confirmed empirical generalizations does not constitute a science. A science is not a catalogue of observed regularities. It requires a certain *rational architectonic*, relating a variety of empirical generalizations in a rational structure that exhibits their conceptual relevance and their explanatory interconnections. A well-established generalization qualifies as a *scientific law* (in the proper sense of the term) only when it finds its theoretical home within some scientific discipline or branch of science.

Yet, as we have seen, the conditions that establish a generalization as *law-like* – that is, as rationally *qualified* for an imputation of lawfulness on the basis of the usual methodological considerations – do not suffice to *establish* it as a law. Because its acceptance as a law involves the claim of lawfulness and the contents of claim extends well beyond the basis upon which it is justified. To class a generalization as *law-like* is to say it is a candidate-law on the basis of factual considerations, but to class it as *lawful* is to step beyond this claim into the realm of nomic necessity and hypothetical force.

Various writers have long argued that the very idea of lawfulness is at bottom anthropomorphic.[15] The basic concept is that lawful phenomena are *rule-governed*: the conception of operative rules is the foundation of lawfulness. But this idea ultimately originates in man's first hand experience of the rules of his social group: rules of behavior, of speech, of dress, etc. The 'pressure' of social rules and the associated sanctions is something of which each man is conscious in his own mind. The rule-conformity of this social context is projected into external nature to provide the concept of lawfulness, analogizing the regularity of social phenomena to those of inert nature, and correspondingly analogizing the alternative range of the socially permissible to that of the naturally possible, and that of the socially obligatory to that of the naturally necessary. In thus viewing man's first-hand experience of social rules as the foundation for his projection of nomic force into the laws of nature, one sees the subjective tendency of mind as the model of lawfulness, the ultimate source for the imputation of nomic necessity that is the touchstone of laws.

In saying that laws are man-made – that they result from a human decision to accord a certain status to specific generalizations – we do not intend to turn our back upon the findings of methodologists of science and theorists of inductive logic. Insofar as their findings conform to the actualities of scientific practice we accept them in full. We are not attempting a Quixotic substitution of 'free decision' for scientific method. But we regard the principles of the theory of scientific method from our own perspective – not as procedures for the *establishment* of generalizations as lawful, but as procedures for providing endeavors a rational warrant for *imputations* of lawfulness.

Returning to the idea of lawfulness as a *well-founded imputation* we remark that both of these two factors, the factual element of well-foundedness and the decisional element of imputation, are necessary to laws. Well-founding is essential because the very spirit of the scientific enterprise demands reliance only upon tested generalizations that have a solid observational or experimental basis. But the element of imputation is also essential. We can only observe what *is*, i.e., forms part of the realm of the actual – not what corresponds to the modally necessary or the hypothetically possible. The nomic necessity and hypothetical force characteristic of lawfulness thus represent factors that a generalization can never earn for itself on the basis of observational or experimental evidence alone. It has to be *endowed* with these factors.

III. LAWFULNESS AS MIND-DEPENDENT

Our conception of the nature of lawfulness carries Kant's Copernican revolution one step further. Hume maintained that faithfulness to the realities of human experience requires us to admit that we cannot find nomic necessity in nature. Kant replied that necessity does indeed not reside in observed nature but in the mind of man, which projects lawfulness into nature in consequence of features indigenous to the workings of the human intellect.[16] Our view of the matter agrees with Hume's that lawfulness is not an observable characteristic of nature, and it agrees with Kant that it is a matter of man's projection. But we do not regard this projection as the result of the (in suitable circumstances) inevitable working of the psychological faculty-structure of the human mind. Rather, we regard it as a matter of *warranted decision*, a deliberate man-made imputation effected in the setting of a particular conceptual scheme regarding the nature of explanatory understanding. We thus arrive at a position that is Kantian with a difference. Kant finds the source of lawfulness in the way in which the mind inherently works. We find its source in the conceptual schemata that we in fact deploy for explanatory purposes: As we see it, lawfulness demands an imputational step made in the context of a certain concept of explanation. Both of these divergent views agree, however, in making lawfulness fundamentally mind-dependent.

On such a view, laws – even natural laws – are man-made. Does it follow from this position that if there were no men – or rather no rational minds – that there would be no natural laws? Are we driven to a law-idealism as the logical terminus of the line of thought we have been tracing out? The answer to these questions, I believe, must be: *Yes*. Given the concept of a law that we actually deploy in these discussions, hypothetical force (and so nomic necessity) is an essential feature of a 'law'. The mode of lawfulness built into the very concept of a natural law involves an essential reference to the domain of supposition and counterfact, to the hypothetical realm of 'what would happen if'. And if rational minds were abolished, the realm of supposition and counterfact would be abolished too, and with it lawfulness as we conceive of it – which involves an essential reference to counterfact – would also have to vanish.

I am not saying simply that laws, being formulated in propositions, presuppose language, and therefore presuppose minds. Nor do I want to go off into Bishop Berkeley's forest. For we are not at the moment concerned with the general idealist position that properties in general require minds. We recognize and admit – indeed regard as crucial – the distinction between the *attribution* of a property to an object by someone (which evidently requires a mind), and the *possession* of the property by the object (which is or may be prepared to be an 'objective', mind-independent fact). But hypothetically counterfactual propositions are mind-related: the hypothetical cannot just 'objectively be' the case, but must be hypothesized, or imagined, or assumed, etc. Unlike real facts, hypothetical ones lack, *ex hypothesi*, that objective foundation in the existential order which alone could render them independent of minds.

Of course, in a trivial sense everything that is discussed bears *some* relationship to a mind. Unquestionably, no matter what truth *we* may think of, *somebody* thinks of: but what people think is not the crux. Being thought of is not essential to the truthfulness of a truth. This whole way of approaching the matter – with reference to what 'is thought' to be the case – loses sight of the key issue of *laws*. This would trivialize the issue: it fails to differentiate laws from generalizations that are not law-like, and so the point loses any specific relevancy to lawfulness as such. But just such reference to lawfulness is the essential thing. I have no desire to question the distinction between a fact, say that the cat is on the mat (which could continue unchanged in a world devoid of intelligence), and the thought or statement of a fact (which could not). My point is that the *claim* of lawfulness, unlike the *claim* of factuality, involves something (viz., a reference to the hypothetical) that would be infeasible in the face of a postulated absence of minds. A generalization like 'All cats are vertebrates (i.e., have backbones)' *if not taken to formulate a law* makes a claim whose correctness is doubtness unaffected if we postulate a mindless universe. But if the generalization is construed in a lawful sense as asserting that cats *have to* have backbones with some sort of nomic necessity, the story is quite different. For lawfulness 'lies in the eyes of the beholder', since the lawfulness of a generalization consists in its being regarded and treated and classified and used in a certain way. All this is impossible in a mindless world. Kant was quite right. Lawfulness is not something that one can meaningfully postulate objectively of a mindless world – it is a mode of 'appearing to a mind'. For if the hypothetical element (which is clearly accessible only in a world endowed with minds) were *aufgehoben* (annihilated), lawfulness would be *aufgehoben* too. Of course, *we* can think of an 'alternative possible world' that is unpopulated, and so mindless, but yet lawful – so long as we do not imagine *ourselves* too wholly out of the picture so as to create a genuinely mindless universe. But if we rigorously put aside all reference to the mental, then the hypothetical possible is lost, and lawfulness is lost with it.

It should be stressed that the hypothetical element at issue here extends well beyond the sphere of overt laws, to encompass all dispositional predicates such as 'soluble', 'malleable', 'fragile', etc., as well as cognate dispositionally classificatory nouns such as conductor (of electricity), nonconductor, etc. For these all have an intrinsic hypothetical element. The cube of sugar is *soluble* because '*if* it is immersed in water for a sufficient period, *then*...'; the copper wire is a conductor because '*if* an electric charge is placed at one extreme, *then*...'. Applications of all such dispositional nouns and predicates are implicitly lawful. Thus our analysis of the implications of lawfulness are operative here also. Thus insofar as lawfulness is mind-dependent so is the applicability of such intrinsically hypothetical qualifiers (unlike purely descriptive qualifiers).[17]

At this point, however, the distinction between laws and regularities becomes important. No doubt nature is in various respect regular – it would take a bold act of rashness to deny that! And this regularity of nature in various respects is no doubt an ontological fact that would remain unaltered in the face of any hypothetical removal of rational minds from within its purview. But the idea of a law involves – as we saw – more than just factual regularity as such, since lawfulness is bound up with nomic necessity and hypothetical force. To say that these factors do not represent objective facts but result from man-made imputations is not to gainsay the objective reality of regularities in nature. Rather, it is to recognize that laws play a role in our conceptual scheme that imposes requirements going beyond mere regularity. It is not the regularity claimed by a law but the lawfulness it builds into this claim that is mind-dependent. (The 'idealistic' aspect of our law-idealism is thus a qualified one.)

The point can be brought home by means of considerations already alluded to. The thesis that 'Oak trees are deciduous' may well represent a regularity that continues unchanged in a mindless world. But the thesis that 'Oak trees *have to* be deciduous' in a sense that warrants 'If that pine were an oak, then it would be deciduous' would not be unchanged. It does not deal with the 'objective fact' of regularity alone but brings in a realm of the hypothetically possible – and this, by the very nature of the hypothetical, is vulnerable to a supposition of mindlessness. Regularity calls for no more than a universal generalization of the type:

(1) All X is Y.

But lawfulness, as we have said, goes beyond this to stipulations of the form:

(2) All *X* is *Y*, and if *z* (which is not an *X*) were an *X*, then *z* would be a *Y*.

Now (1) is simply an issue of ontological fact: if it is in fact the case that *All (pure) mercury solidifies at* $-150\,°$F, then this circumstance could continue operative even if there were no minds around to think about such matters. But a type (2) generalization – a lawful generalization – would fare differently. Its reference to *z*'s which are not *X*'s having to have certain characteristics if they were *X*'s involves claims outside the domain of ontological fact, claims of which sense can be made only under the assumption that there are minds somewhere upon the scene, capable of entertaining hypotheses regarding what-would-be-if. In a mindless universe, the whole domain of the hypothetical can find no foothold. This, then, is the foundation for our thesis that laws, involving as they do essentially hypothetical claims, are mind-dependent in a way that endows them with an inevitable man-made component.

The critical point that the realm of the hypothetical is mind-dependent should be argued explicitly. Somewhat reluctantly we must thus enter upon a metaphysical digression regarding the ontology of the possible. The argument for the mind-dependency of hypothetical possibilities proceeds as follows:

(1) The natural world comprises only the actual. This world does not contain a region where nonexistent or unactualized possibilities somehow 'exist'. Unactualized hypothetical possibilities do not exist in the world of objective reality at all.

(2) Nor do unactualized possibilities exist in some Platonic realm of world-independent reality.

(3) The very foundation for the distinction between something actual and something merely hypothetically possible is lacking in a 'mindless' world. Unactualized hypothetical possibilities can be said to 'exist' only insofar as they are *conceived* or *thought of* or *hypothesized* and the like. For such a possibility to be (*esse*) is to be conceived (*concipi*).[18]

In such a way, then, one can argue a denial that possibilities exist in some self-subsisting realm that is 'independent of the mind'. Inorganic nature – subrational nature generally – encompasses only the actual: the domain of the possible is the creation of intelligent organisms. A 'robust realism of physical objects' is all very well, but it just will not plausibly extend into the area of the hypothetical. It would be foolish (or philosophically perverse) to deny the thesis: 'This (real) stone I am looking at would exist even if nobody saw it'. But we cannot reason by analogy to support the thesis: 'This imaginary stone I am thinking of would exist even if nobody imagined it'. The objectivity of the real world does not underwrite that of the sphere of hypothetical possibility. This sphere is mind-dependent, and so consequently are those intelligence presupposing resources which – like laws – are hinged upon it.

Of course, Idealists of the old school (Bradley, Blanshard) would not accept our conception of laws as a *via media* between mere regularities on the one hand and the logically necessary on the other. Indeed their reasoning seems to proceed by elimination: there is no such intermediary and natural laws are not mere regularities, *ergo* they are truths of logic. Now, of course, one would incline to counter this position with the claim that there is nothing inherently self-contradictory about denying the laws of nature while there is something self-contradictory about denying a logical truth. The old-line idealists will respond that low-denials will turn out to be self-contradictory once we have learned enough about the *system* of which they are apart. (It is certainly possible to hold a belief on grounds one believes to be empirical, but which later turns out to be logically necessary.) They hold that when science is complete, that all lawful relations will be revealed as somehow logically necessary. This (Blanshard's) position is akin to that of Leibniz: All empirical propositions are analytic although they will not appear so to our finite minds. This position is not patently untenable, but it places a burden on one's faith in ultimates that has little to recommend it.

The key points of the argument developed in these three sections can be summarized as follows:

(1) The concept of scientific explanation is such as to require *lawfulness* in the generalizations employed.

(2) Lawfulness requires the factors of nomic necessity and hypothetical force.

(3) Nomic necessity and hypothetical force both in significant measure go beyond the sphere of what can be established by observation and experiment.

(4) Lawfulness can thus never be wholly based upon an observational foundation. Rather, it represents an *imputation* that is (or should be) well-founded upon evidential grounds. (The key factors in this well-foundedness are the *correspondence-to-fact* aspect of empirical evidence and the *systematic-coherence* of filling the generalization into a fabric of others that in the aggregate constitute a rational structure, an integrated body of knowledge that constitutes a 'branch of science'.)

(5) Laws are therefore in significant respects not discovered, but made. A law, unlike a simple assertion of regularity, involves claims (viz., of nomic necessity and hypothetical force) that are mind-dependent and cannot be rested simply upon objective matters of observed fact.

(6) Our position thus has the character of a qualified idealism. Lawfulness is not *just* a matter of the observable facts, but involves – through reference to the factors of nomic necessity and hypothetical force – an essential element of transfactual imputation, and thus is in a crucial respect mind-dependent.[19]

University of Pittsburgh

REFERENCES

[1] It is clear that we mean this to be construed as 'if it were somehow *added to* the coins in my pocket' and not as 'if it were to be somehow *identical with* one of the coins in my pocket'.

[2] Roderick M. Chisholm has put this point with admirable precision: 'Both law statements and non-law statements may be expressed in the general form, "For every *x*, if *x* is an *S*, *x* is a *P*". Law statements unlike non-law statements, seem "however" to warrant inference to statements of the form, "If *a*, which is not *S*, were *S*, *a* would be *P*" and "For every *x*, if *x* were *S*, *x* would be *P*"'. R. M. Chisholm, 'Law Statements and Counterfactual Inference', *Analysis* **15** (1955), 97.

[3] The considerations at issue here are treated in more detail in my book on *Hypothetical Reasoning*, North-Holland Publ. Co., Amsterdam, 1965.

[4] See especially Secs. 1–6 of Bk I.

[5] See especially the Introduction and the Transcendental Analytic.

[6] See Curt J. Ducasse, 'Explanation, Mechanism, and Teleology', *The Journal of Philosophy* **23** (1926), 150–155. Reprinted in H. Feigl and W. Sellars (eds.), *Readings in Philosophical Analysis*, Appleton Century Crofts, New York, 1949. Arthur Pap, *An Introduction to the Philosophy of Science*, The Free Press, New York, 1962, Chapter 16.

[7] R. M. Chisholm, 'The Contrary-to-Fact Conditional', *Mind* **55** (1946), 289–307, reprinted in H. Feigl and W. Sellars, *Readings in Philosophical Analysis*, *op. cit.*; Nelson Goodman, 'The Problems of Counterfactual Conditionals', *The Journal of Philosophy* **44** (1947), 113–128, reprinted in L. Linsky (ed.), *Semantics and the Philos-*

ophy of Language, University of Illinois Press, Urbana, Ill., 1952, and in N. Goodman, *Fact, Fiction, and Forecast*, Harvard University Press, Cambridge, Mass., 1955.

[8] R. S. Walters, 'Laws of Science and Law-like Statements' in the *Encyclopedia of Philosophy* (ed. by Paul Edwards), vol. IV, Macmillan, New York, 1967, pp. 410–414 (see pp. 411–412). This article offers a very clear and compact survey of the key issues regarding laws.

[9] Mario Bunge, *The Myth of Simplicity*, Prentice-Hall, Englewood Cliffs, 1963, p. 174. At least one influential adherent of the regularity theory was, however, prepared to brush aside all reference to the possible, saying: 'Physics wants to establish regularities; it does not look for what is possible'. (L. Wittgenstein in his middle period as quoted by H. Spiegelberg in the *American Philosophical Quarterly* **5** (1968), 256.)

[10] R. S. Walters, *op. cit.*, pp. 413–414.

[11] This position is in fact taken by latter day idealists of the type of Bradley and Blanshard who hold in effect that nomic necessity and logical necessity are indistinguishably one and the same, so that counterfactual hypothesis cannot be posed at all in any meaningful or coherent way.

[12] For a cogent attack on the view that laws can be established by induction see K. R. Popper, *The Logic of Scientific Discovery*, Hutchinson and Co., London, 1959, Chapter III and New Appendix 10.

[13] It is obviously naive to think that one can settle the question of the *counterfactual* application 'If Caesar's chariot had been a satellite in orbit about the earth it would have moved according to Kepler's laws' by increasing the domain of *actual* applications of Kepler's laws by putting more spacecraft into orbit.

[14] See for example, the excellent discussion in Chapter 4 of Ernest Nagel's book on *The Structure of Science*, The Free Press, New York, 1961.

[15] See Edgar Zilsel, 'The Genesis of the Concept of Physical Law', *The Philosophical Review* **51** (1942), 3–14.

[16] A thread running constant throughout the history of philosophy is the thesis that there would be no laws if there were no lawgiver: that the universe would not be intelligible by man if it were not the product of a creative intelligence. We find this theme in Plato's *Timaeus*, in the cosmological argument of St. Thomas Aquinas and the schoolmen, in Descartes and Leibniz, in Butler's *Analogy* and the tradition of natural theology in England, etc. Leibniz puts the matter cogently and succinctly: '... the final analysis of the laws of nature leads us to the most sublime principles of order and prefection, which indicate that the universe is the effect of a universal intelligent power'. G. W. Leibniz, *Philosophical Papers and Letters* (ed. by L. E. Loemker), vol. II, University of Chicago Press, Chicago, 1956, pp. 777–778. Kant in effect agrees with the underlying thesis that the intelligibility and rationality of the universe must be the work of an intelligent and rational mind, but shifts the application of the principle from the *creator* of the natural universe to the *observer* of it.

[17] Someone might contend that ordinary, overtly nondispositional predicates like 'has a length of one meter' are all covertly dispositional in the manner of: 'If one were to take a meter rod and lay it alongside, then ...'. This assimilation of all physical predicates to the dispositional ones, though superficially plausible, is at root indefensible. To be sure, all such descriptive predicates have a regularity component and a lawfulness component; but with ordinary 'descriptive' predicates, the regularity component predominates, while with dispositional predicates the lawfulness component is predominant. The straightforward assimilation of the two cases is thus an unjustified oversimplification.

[18] To say this is not to drop the usual distinction between a thought and its object. If I imagine this orange to be an apple, I imagine it *as an apple*, and not as an *imaginary* apple. But this does not gainsay the fact that the apple at issue 'exists only in my imagination'.

[19] Some of the ideas dealt with in the present section are discussed in a very illuminating way in A.C. Ewing, *Idealism: A Critical Survey*, Methuene and Co., London, 1934, 3rd ed., 1961, see chap. viii, 'Idealistic Metaphysics'. The materials of this present essay are placed in their wider systematic context in N. Rescher, *Scientific Explanation*, The Free Press, New York, 1969.

JAEGWON KIM

EVENTS AND THEIR DESCRIPTIONS: SOME CONSIDERATIONS

Events presumably are not linguistic entities; like trees and molecules, events can be talked about, referred to, and described but they are not themselves statements, sentences, descriptions, or any other kind of linguistic units. Nor are events propositions; propositions are supposed to be abstract entities, whereas events are spatio-temporally bounded particulars. Events and relations between events would exist even if there were no humans, or language, to describe them; there still would be earthquakes, collisions of particles, and expansions of metals caused by rising temperatures.

All this is surely obvious and trivial. Moreover, what has been said about events is equally true of such related categories as *states*, *conditions*, *processes*, and also perhaps *facts*. But consider the following passages taken from some recent philosophical writings:

> In formal language, an event Y at the time T is caused by a preceding event X, if and only if Y is *deducible* from X with the aid of the laws L_T known at the time T.[1]

> ... all that is important here is the recognition insisted upon by Hume that natural events (e.g., explosions, cell divisions, etc.) which are causally related are *logically independent* of one another.[2]

> A type of explanation commonly encountered in the natural sciences, though not exclusively in those disciplines, has the formal structure of a deductive argument, in which *the explicandum is a logically necessary consequence of the explanatory premises....*
> In both the second and third examples, the *explicandum* is a historical *fact*. However, in the second the fact is an individual *event*, while in the third, it is a statistical *phenomenon* ... .[3]

None of these passages makes clear sense unless one can meaningfully speak of the 'deduction' of an event from another event or from statements. The relation of logical deducibility applies only to sentences, statements, or propositions (and perhaps sentence-forms, etc.); one can no more deduce an event or deduce something from an event than deduce a tree or table from something or deduce something from a tree or table. Further, if events are not proper terms for the relation of deducibility, they cannot be proper terms for the relation of logical implication. Thus

N. Rescher et al. (eds.), Essays in Honor of Carl G. Hempel.

it is meaningless, in the absence of further explanation, to speak of the logical dependence or independence among events. In the last of the quoted passages, the term 'explicandum' does a double duty for both an event and a descriptive statement standing for that event. This ambiguity serves as a reminder that there exists very little discussion in the literature on an adequate analysis of the concept of event or the concept of description of an event.[4] This is a surprising fact in view of the central role the concept of event plays in some of the more important problems in philosophy today, such as causation, determinism, the nature of human action, and explanation. And philosophers who customarily use the notion of 'description of an event' (e.g., philosophers who use the locution 'event (or action) under a description'[5]) have done little to clarify it. As we shall see, this notion is far from being clear or self-evident; in fact it gives rise to some puzzling problems, and it is not at all certain that the concept can be elucidated in a precise and fruitful way.

In what follows, I shall first try to motivate the problem of explicating 'description of an event' in connection with the theory of explanation and causation. I shall then state and examine what seems to be a simple and intuitive approach to an analysis of this notion; I shall be particularly concerned with the viability of this approach in face of an argument derived from Frege and Quine which seems, at first blush, to constitute a devastating objection against it.

I

According to the so-called 'covering-law model' or 'deductive nomological model' ('D-N' for 'deductive-nomological' hereafter) of explanation,[6] an explanation of an individual event is a deductive argument with general laws and singular statements of initial conditions as premisses and a singular statement as its conclusion. Following Hempel's terminology, we refer to the premisses of such an argument as the *explanans* and to its conclusion as the *explanandum statement*; the event thus explained is called the *explanandum*. The use of general laws in the explanans is thought to be essential, whence the further requirement that the explanandum statement be not deducible from the singular premisses of the explanans alone. A 'true' or 'correct' explanation must have a true explanans; that is, every statement in the explanans of a true explanation must be true. When an argument is exactly like an explanation except possibly that its explanans is false, we call it a *potential* explanation; thus, an explanation is a potential explanation with a true explanans. Let us now summarize this:

(DN) The ordered pair (L, C) of law-like statements and singular statements of initial conditions is a *potential deductive explanans* for the singular statement E if and only if E is a deductive consequence of L and C but not of C alone.[7]

Since E is a statement (or a sentence or proposition), the relation 'is an explanans for' as defined by (DN) is a relation between statements and not between a set of statements and an event; that is, it does not relate the explanans directly to the event being explained. On the other hand, we normally think of an explanation as an explanation *of an event* (or a *state* or *fact*, and so on) and not of such things as statements or sentences, unless of course the explanation specifically concerns, e.g., the meaning of a sentence.

However, it will not do to restate (DN) simply by substituting 'event e' for 'E', thus:

(1) (L, C) is a potential deductive explanans for event e if and only if e is a deductive consequence of L and C but not of C alone.[8]

For no clear sense can be attached to the idea that an *event* can be a logical consequence of some *statements*. We cannot say, therefore, that (DN) as stated gives a complete account of deductive explanation of individual events. What is lacking is a provision connecting the logical relation defined by (DN) between statements with the explanatory relation which, intuitively, relates sets of statements to the actual events and facts of the world. (DN) tells us the precise conditions under which an argument is an explanatory argument; what it fails to tell us is which particular event or fact a given explanation is an explanation of. So, with (DN) alone, we would be in a position in which we are able to decide when we have an explanation and when we do not but unable to decide which events our explanations purport to explain.

In order to remedy this deficiency one naturally looks for some kind of semantical relation that would relate the explanandum statement to the event being explained. Thus, Hempel writes:

We divide an explanation into two major constituents, the *explanandum* and the *explanans*. By the explanandum, we understand the sentence describing the phenomenon to be explained .... [9]

Although the word 'describe' is most often used to denote this relation, others are also used, such as 'refer to', 'specify', 'represent', and 'assert the occurrence of'. We shall mainly use 'describe' in this paper but on occasion the others will also be used interchangeably. Precisely how this notion of describing an event is to be explicated is the major problem that will occupy us in the remainder of the paper; but let us first see how, by the use of this notion, the relation of 'is an explanans for' is to be characterized in order to relate an explanans directly to the event being explained. One might first try something like the following formulation:

(2) Let E describe e. Then (L, C) is an explanans for e if and only if (L, C) is an explanans for E by (DN).[10]

This statement, however, leads to an overly stringent condition on the identity of events described by statements. Given the unexceptionable principle that if event e is identical with event e', any explanans for e is an explanans for e', it follows from (2) that two statements describe the same event only if they are logically equivalent.[11] As a particular example, on the assumption that 'Anyone who drinks hemlock dies' is a law, it follows from (2) that 'Socrates died' and 'Xantippe's husband died' describe different events; for ('Anyone who drinks hemlock dies', 'Socrates drank hemlock') is an explanans for the former statement but not the latter.

A familiar type of analysis of the concept of causation attempts to explain the relation of causation between two events in terms of a logical relation between statements describing them. The following formulation, which is in the spirit of (2), leads to much the same consequences as does (2):

(3) Let E and E' describe events e and e' respectively. Then e causes e' if and only if there exists a (causal) law L such that E' is a logical consequence of L and E but not of E alone.[12]

If we assume that 'Anyone who drinks hemlock dies' is a causal law, it follows that the event described by 'Socrates drank hemlock' did not cause the event described by 'Xantippe's husband died', although of course it did cause that described by 'Socrates died'; it further follows that the event described by 'Socrates died' is not the same as that described by 'Xantippe's husband died'.

I take it that these are unacceptable results; if any useful meaning is to be attached to the talk of a statement's describing an event or phenomenon, it seems clear that one and the same event is described by 'Socrates died' and 'Xantippe's husband died'. Not only is this intuitively correct but no useful theoretical purpose seems to be served by denying it and thereby multiplying events. I have drawn attention to (2) and (3) not because of their intrinsic interest but only as a way of drawing attention to some aspects of our problem. In any case, (2) and (3) are easily revised so as to avoid the unwanted multiplication of events.

(DN′) (L, C) is an explanans for event e if and only if there is a singular statement E such that E describes e, and (L, C) is an explanans for E by (DN).[13]

Under this revised formulation we can say that ('Anyone who drinks hemlock dies', 'Socrates drank hemlock') is an explanans for the event of the death of Xantippe's husband, for we can say that 'Socrates died' describes the death of Xantippe's husband. Further, unlike (2), (DN′) does not have the consequence that the event described by 'Socrates died' and that described by 'Xantippe's husband died' are distinct events, for we can say that the above ordered pair of statements is an explanans for the event described by 'Xantippe's husband died' in virtue of the fact that this event is the very same event that is described by 'Socrates died'.

Since (3) too led to the unwanted consequence that 'Socrates died,' and 'Xantippe's husband died' specify distinct events, we need to restate it:

(4) Event e caused event e' if and only if there are singular statements E and E' such that E and E' describe e and e' respectively, and E' is a logical consequence of E together with some causal law L but not of E alone.[14]

Evidently, (4) does not have the consequence that the event described by 'Socrates drank hemlock' did not cause the event described by 'Xantippe's husband died'; on the contrary, under the assumption that 'Socrates died' and 'Xantippe's husband died' describe the same event, we have the result that the event described by 'Socrates drank hemlock' did cause the event described by 'Xantippe's husband died'.

The main problem, however, is to construct a general criterion which states the conditions under which two event-describing statements describe the same event. This of course is not the only, or even the most important, problem about the concepts of event and event-description; one important set of problems concerns the ontological and logical nature of events; for example, one such question is whether or not an event can be analyzed in terms of 'simpler entities' such as substances (or bodies) and properties; another problem in this category will concern the interconnections and dependencies between such related categories of entities as events, processes, states, and facts. These are large questions which lie outside the scope of this paper. We shall also eschew the general problem about event-description, namely that of giving a truth condition for 'Statement E describes event e' and confine our attention to an analysis of 'Statement E describes the same event as statement E''. One important reason for narrowing the scope of our inquiry in this way is that up to a point this last question can be dealt with more or less independently of the questions concerning the ontological and logical nature of events.

This problem concerning the conditions under which two statements describe the same event can be motivated in another way. Consider the locution 'The explanans T explains why...', where the dotted line is a place-holder for singular event-describing statements; thus, a particular instance of it will be '('Anyone who drinks hemlock dies', 'Socrates drank hemlock') explains why Socrates died'. The problem I have in mind is what kinds of substitution, if any, are permissible in the dotted context. If we accept the principle that if event e is the same event as event e' any explanans of e is an explanans for e', then it follows that the replacement of the statement occupying the dotted line by another describing the same event will not affect the truth of the whole statement. For example, the replacement of the second occurrence of 'Socrates' by 'Xantippe's husband' in '('Anyone who drinks hemlock dies', 'Socrates drank hemlock') explains why Socrates died' yields the true statement '('Anyone who drinks hemlock dies', 'Socrates drank hemlock') explains why Xantippe's husband died'. The relation between this problem and the problem concerning the identity of events described by two statements is obvious: The permissible logical operations in the dotted context of 'T

explains why...' are exactly those which, when performed on an event-describing statement, will leave intact the identity of the event described by it.

II

In order to forestall certain objections which are largely irrelevant to my purposes in this paper, I must first make a few qualifications. First, I am not here concerned with the various possible distinctions one might wish to make among the entities I call 'events'; I am using this blanket term to comprehend not only what we ordinarily call 'events' but also such entities as 'states', 'states of affairs', 'phenomena', 'conditions', and the like. Perhaps 'fact' is more appropriate, although it too has its disadvantages. This extremely broad use of the term 'event' is not unwarranted by the customary philosophical usage; in fact, it occurs when we discuss explanation and causation. When we talk of explaining an event, we are not excluding what, in a narrower sense of the term, is not an event but rather a state or process, e.g., the sky's being blue and the growth and spread of cancer. Similarly, when we discuss the problem of causation in terms of events, we are not thereby implying that causal relations cannot properly obtain between processes, states, and conditions. In short, I am interested in events qua objects of explanation and relata of causal relation; I choose the term 'event' because that is the term we most often use when discussing explanation and causation.

Second, by 'a description of an event' I have in mind what might more properly be called a *sentential* or *propositional description* of an event; thus, logically, the linguistic units in which I am interested are statements, and not nominal phrases such as definite descriptions (e.g., 'John Smith's most unforgettable event') purporting to make unique references to individual events.[15] Moreover, the kind of statement in question will generally not include any such referring expressions for events; our paradigmatic event-describing statements are statements like 'Socrates drank hemlock' and 'Xantippe wept'. The idea is that the statement as a whole, not any proper part of it, specifies or describes a concrete event or state. These are typically the sort of statement which, in discussions of explanation and causation, are said to 'describe events'. In any case, these statements can serve as explanandum statements in 'explanations of events'; it is quite in order to speak of explaining why Socrates drank hemlock or why the Titanic sank. In explaining them we would be explaining events. We can also inquire into the causes of these events. Thus, if we are told that the Titanic sank in 1912, we might ask what caused this *event*, and how this *event* happened, and so on. And the event we are here talking about is the one specified by the statement 'The Titanic sank in 1912'. Generally speaking, the class of statements in question can be delineated very roughly by saying that they attribute an empirical property (or relation) to a concrete object (or ordered set of objects) at a time. And we assume that for each such event-describing statement there is some unique event specified or described by it; this assumption is needed to warrant our talk of *the* event described by a given statement.

Consider again 'Socrates died' and 'Xantippe's husband died'. The reason why we are inclined to think that there is one event, and not two, that is given by the two statements appears to be that, intuitively, the statements are 'about' the same object, namely Socrates, and assert of this object that the same property holds for it. There is a particular object, Socrates, of which the statements affirm that it died, and as far as reference to this event is concerned, it should make no difference whether we refer to the object as Socrates or Xantippe's husband. From a logical point of view, the idea is that if we replace a term by another co-designative term in an event-describing statement, the resulting statement describes the same event. This is the idea that I want to explore and examine in the remainder of this paper. Its ultimate usefulness depends on whether it can plausibly be generalized; as we shall see, serious difficulties arise in constructing a coherent general account of event-description in this direction. But the idea has some measure of initial plausibility, and is well worth a close scrutiny.

If, however, we are to pursue this line of approach the further question arises at once: Should we also say that if the same property is ascribed by two statements to the same object, although the property is specified by different predicative expressions, the two statements specify the same event? For example, consider 'This pencil is blue' and 'This pencil has the color of the sky'. It would seem consonant with this general approach to say that these two statements specify the same event (state). Events are nonlinguistic entities, and it should not matter what particular terms we use to pick out the objects and properties as long as the same ones are picked out. If some process caused this pencil to turn blue, then it caused it to have the color of the sky for the obvious reason that blue is the color of the sky.

Thus, the idea that in a statement any referring expression can be replaced by another co-referential one without changing the identity of the event described by the statement should in consistency be extended to properties by construing predicative expressions, e.g., 'is red', 'turns red', etc., as referring expressions for properties.[16] Let us note here that, contrary to what some philosophers maintain, it is not necessary for two predicate terms to be synonymous or logically equivalent for them to refer to, or express, the same property; for example, 'is blue' and 'has the color of the sky' clearly show this. In any case, the leading idea is that a statement of the kind in which we are interested is construed as being *about* those objects and properties referred to by the referring expressions occurring nonvacuously in the statement, and that the import of the statement is to affirm of these objects that the properties in question hold for them, or are exemplified by them. Thus, this leads to the following:

(I) If a statement S' is obtained from a statement S by replacing any referring expression in S by a co-referential expression, then S and S' describe the same event.[17]

This principle licenses us to say that 'Socrates died' and 'Xantippe's husband died' describe the same event, and similarly for 'This pencil turned blue' and 'This pencil turned the color of the sky'. However, it does not allow us to say that 'Brutus killed Caesar' and 'Caesar was killed by Brutus' describe the same event; nor does it allow us to say that 'Brutus killed Caesar' and 'Calpurnia's husband was killed by Brutus' describe the same event. In order to take account of these intuitive cases of event identity, a natural and plausible solution would be to adopt the following principle, in addition to (I):

(L) Logically equivalent statements describe the same event.

It is easy to see that repeated applications of (I) and (L) yield the result that 'Socrates died', 'Xantippe's husband died', and 'Xantippe became a widow' all describe the same event, and similarly for 'Brutus killed Caesar', 'Caesar was killed by Brutus', and 'Calpurnia's husband was killed by Brutus'.

The two criteria together state a sufficient condition for statements describing the same event: S and S' describe the same event if one is obtainable from the other by a finite sequence of transformations in accordance with (I) and (L), namely substitutions of identicals and logically equivalent transformations.

Let us now begin a critical examination of the two principles (I) and (L). First, it can be shown that any two true statements of the form 'Object a is F' and 'Object b is G' are such that the event described by one is identical with the event described by a statement which entails the other statement. The paradoxicality of this situation is highlighted if the following natural generalization of (L) is accepted:

(5) If S implies S', the event described by S *includes* the event described by S'; if the event described by S and that described by S' include each other, S and S' describe the same event.

On this definition of 'inclusion' for events, one can show that any two true statements 'a is F' and 'b is G' describe the same event. Consider the following array of statements:

(A) a is F.
(B) b is G.
(C) $(\iota x)\,(x = a\ \&\ b \text{ is } G)$ is F.
(D) $(\iota x)\,(x = b\ \&\ a \text{ is } F)$ is G.

Notice, first of all, that $a = (\iota x)\,(x = a\&b \text{ is } G)$, and similarly that $b = (\iota x)\,(x = b\&a \text{ is } F)$. Thus, by (I), (A) and (C) describe the same event, and (B) and (D) describe the same event. But (C) implies (B) and (D) implies (A). Thus, the event described by (A), by (5), includes, and is included by, the event described by (B); and, by (5), (A) and (B) describe the same event.

The effect of (I) and (L) on the behavior of the context 'T explains why...' is this: according to (I), the dotted context is open to substitution of identicals namely, 'referentially transparent', and according to (I), the context admits of interchange of logical equivalents – at least the interchange of the whole context by a logical equivalent. But this is disastrous. For there is a proof of Quine's[18] showing that any context having these two properties is also fully truth-functional, and clearly we do not wish to make the context of explanation truth-functional, for the obvious reason that it would enable us to explain everything under the sun by a single explanation. More directly, this proof can be used to show that (I) and (L) have the catastrophic consequence that all true event-describing statements describe the same event and that all false event-describing statements also describe the same event (presumably, none) as follows:[19] let p and q be any arbitrary event-describing statements alike in truth-value, and consider the following array of statements:

(a) p
(b) $\{x \mid x = \emptyset \,\&\, p\} = \{\emptyset\}$
(c) $\{x \mid x = \emptyset \,\&\, q\} = \{\emptyset\}$
(d) q

Since (a) and (b) are logically equivalent, they describe the same event, by (L); since p and q have the same truth-value, $\{x \mid x = \emptyset \,\&\, p\} = \{x \mid x = \emptyset \,\&\, q\}$; so by (I), (b) and (c) describe the same event; further, (c) and (d) describe the same event by virtue of their logical equivalence. Thus, it follows that p and q describe the same event. Since false statements presumably describe no event, it would appear that there is one single event that is described by all true event-describing statements, which surely is an intolerable result for any theory of events.

III

Thus, the unrestricted use of both (I) and (L) leads to disaster. In this section I shall examine these two principles individually, show that even their individual uses lead to seemingly objectionable results, and discuss some possible ways of avoiding them. It is clear that if (L) turns out to be untenable any general account of event-description would have to be enormously complicated, and that the whole D-N account of explanation and causation would have to be rethought in light of this fact. For if two statements E and E' are logically equivalent, then under the usual D-N model (or any other simple logico-inferential model) of explanation they are indiscernible as explanandum statements; that is, any explanans for one of the two will be an explanans for the other. But if E and E' describe distinct events, as could happen if (L) is not universally true, a given explanans for E may not be one for E'. I shall later provide some examples which seem to illustrate this point; but I shall not in this paper touch on the possible restatement or reorientation of the D-N account of explanation and causation.

Let us first consider (L). In spite of its initial plausibility and theoretical desirability, this principle does not seem to be in complete harmony with our intuition about events. Our intuition about the event a given statement specifies in our intended sense has much to do with the subject-predicate structure of the statement: the object the statement is *about* is picked out by the subject term, and the predicate term picks out the property (state, process, etc.) the statement affirms of it. This explains why we take it for granted that 'Socrates died' and 'Xantippe's husband died' specify the same event; it explains why we are reluctant to say that certain forms of truth-functionally compounded statements (e.g., conditionals, long disjunctions) describe a 'single' event; I think it also explains why we can think of an explanation of why Wilbur *behaved so obnoxiously* at the party as an explanation of why Wilbur *behaved the way he did* at the party. On the other hand, 'Xantippe's husband died' and 'Xantippe became a widow', though logically equivalent, do not pick out the same subject or the same property. In general, transformations preserving logical equivalence cannot be counted on to preserve the particular subject-predicate structure of a statement, and for this reason it would not be surprising if such transformations failed to preserve the identity of the event described by a statement. In fact, it would not be surprising if such transformations took a statement out of the class of event-describing statements. A case in point is provided by the trivialization argument above: 'Socrates drank hemlock' and '$\{x \mid x=\emptyset$ & Socrates drank hemlock$\}=\{\emptyset\}$'. In spite of their manifest logical equivalence, I am tempted to say that the second statement does not describe any event at all, although the first clearly does. Granted the second statement affirms a certain set-theoretical fact which obtains if and only if the event described by the first exists; but it itself does not describe, refer to, or specify any event. For example, compare: 'Explain why Socrates drank hemlock' and 'Explain why $\{x \mid x=\emptyset$ & Socrates drank hemlock$\}=\{\emptyset\}$', or 'What caused it to be the case that Socrates drank hemlock?' and 'What caused it to be the case that $\{x \mid x=\emptyset$ & Socrates drank hemlock$\}=\{\emptyset\}$?'

Thus, the class of event-describing statements is not closed under logically equivalent transformation, and the scope of the principle (L) would have to be restricted to event-describing statements. If this is done, the trivialization argument breaks down: from the logical equivalence of (a) and (b) it does not follow that they describe the same event, and likewise for (c) and (d). But taking this way out requires a reasonably precise specification of the concept of event-describing statement. What one would want is some canonical scheme for event-description, a standard linguistic form in which event-describing statements would be formulated. In fact, we have already adumbrated such a scheme when we said that the kind of statement in which we are interested is one that attributes a factual property (or relation) to a concrete object (or an ordered *n*-tuple of concrete objects) at a time; a more detailed study of this problem, however, cannot be attempted here.

The restriction of (L) to event-describing statements may not be enough, however. For consider again 'Xantippe's husband died' and 'Xantippe became a widow'. Intuitively, they are both event-describing, but do they describe the same event? Consider their spatial locations: it is plausible to say that the event described by the first statement occurred in the prison in which Socrates took hemlock, whereas the event specified by the second, that is, the event of Xantippe's becoming a widow, occurred not in the prison but in Socrates' home or wherever Xantippe was at the time of Socrates' death.[20] Another possible example is this: 'Wilbur married Edith' and 'Edith married Wilbur'. They are logically equivalent, but to explain why Wilbur married Edith is not necessarily the same as explaining why Edith married Wilbur. Further, we would normally give different answers to the two questions: 'What caused Wilbur to marry Edith?' and 'What caused Edith to marry Wilbur?' Thus, from both the explanatory and the causal point of view there is reason to think that there are two events here, not one.[21] It should be obvious that if these are genuine counter-examples to (L), they present a fundamental difficulty to the D-N model of explanation and causation, or *any* other simple logico-inferential account of explanation and causation.

Let us now turn to (I). In contrast to (L), the type of transformation sanctioned by (I) will in general preserve the subject-predicate structure of a statement, and in consequence, the object an event-describing statement is about and the property it attributes to it would appear to remain invariant from statement to statement as long as (I) alone is applied. But some apparent difficulties for (I) are easily generated. Consider Wilbur, the black crow: since black is the color of Wilbur's plumage, it follows from (I) that 'Wilbur is black' and 'Wilbur has the color of Wilbur' describe the same event (state).[22] But is this correct? Consider another example: our man Wilbur is both the best player and the fastest man on the team. Then, again by (I), it follows that 'Wilbur is the fastest man on the team' and 'Wilbur is the best player on the team' describe the same event (state). But this seems just wrong; to cite one reason, there is an obvious explanatory, and perhaps also causal, asymmetry here. We can explain why Wilbur is the best player by pointing to the fact that he is the fastest man, but hardly the other way around. What is worse, it follows, further, that these two statements specify the same event (state) as does 'Wilbur is Wilbur'.

These objections, however, do not seem fatal to the criterion (I). I shall indicate here, without detailed discussion and justification, a possible line of defense. The best way of meeting the second of the two difficulties above seems to be to argue that 'Wilbur is the fastest man on the team' and 'Wilbur is the best player on the team' are not genuine identity statements but rather statements of predication, and that 'the best player on the team' and 'the fastest man on the team' are not used to make 'genuine reference'. The distinction I have in mind between genuine identity statements and identity statements which are in effect statements of predication is that of David Wiggins.[23] According to Wiggins, not every statement which consists of the identity sign flanked by two noun phrases is a genuine identity statement; for it to be one the two noun phrases must serve 'independently of one another to make genuine references', where 'genuine reference' is explained as follows:

> A speaker refers to something by an expression *e* if (i) he is in a position (and thinks he is) to accept and answer informatively the question 'Which particular are you identifying by *e*?' and (ii) he is in a position to answer this question without making use of the fact that his statement is true, and (iii) he knows (and thinks he knows) how to locate (directly or indirectly) the item which he means to identity by *e* (p. 45).

In this sense, 'The morning star = the evening star' counts as an identity statement whereas 'Wilbur is the fastest man on the team' and 'Wilbur is the best player on the team', used in a normal context, count not as identity statements but as statements of predication. In the latter two statements, 'Wilbur' is used to make genuine reference, but 'the fastest man on the team' and 'the best player on the team' are not so used; rather, these two expressions are used to pick out the properties, being faster than anyone else on the team and being a better player than anyone else on the team, which the statements affirm of Wilbur. On the other hand, both the expressions flanking '=' in 'The morning star=the evening star' are used, independently of each other, to pick out objects, namely the star of such and such sort visible in the morning and the star of such and such sort visible in the evening, and the statement affirms of these objects that they are one and the same.

It is not certain that the concepts of genuine identity statement and genuine reference can ultimately be made clear and defended, but let us assume that they can. We should then introduce a restriction on (I) to the effect that any concrete singular referring expression to be replaced by a co-referential one in accordance with (I) must be used to make genuine reference in the statement to which (I) is applied. We should also exclude all genuine identity statements from the class of event-describing statements; this policy ties in with our earlier suggestion that an event-describing statement is one that attributes an empirical property to an object at a time. Under these revisions, (I) no longer yields the unwanted result that 'Wilbur is the fastest man on the team' and 'Wilbur is the best player on the team' describe the same event (state). This would be possible only if the property of being the best player on the team and the property of being the fastest man on the team were the same property, which presumably is not the case.

The exclusion of genuine identity statements from event-describing statements is not as *ad hoc* as it might seem. Although it cannot be argued in detail here, these statements, unlike those identity statements which are statements of predication, cannot properly function as explanandum statements, i.e., the conclusions of explanatory arguments. It makes no sense to ask for an explanation of *why* the evening star is identical with the morning star; nor does it make sense to talk of the *cause* that brought it about, or made it happen, that the evening star is identical with the morning star. Our proposal explains this fact by pointing out that there is no event or state which is given by the statement 'The evening star=the morning star'. On the other hand, it makes eminently good sense to explain why Wilbur is the best player on the team or why McCarthy was the winner in the Wisconsin primary, or speak of what caused or brought about these states of affairs.

Now for the difficulty raised by 'Wilbur is black' and 'Wilbur has the color he has': the best way of handling it seems to be to accept the consequence that the two statements describe the same event (state). There are conceivable situations in which the second statement can be used to make the same assertion, to point to the same fact or situation, as the first; we do say things like 'Wilbur is the way he is because of his wife', 'Things being as they are...', and so on.[24] Moreover, we can explain why Wilbur has the color he has. We shall have given an explanation of this event (state) when we have explained why Wilbur is black, for black *is* Wilbur's color. It is unlikely that we can construct a D-N explanation conforming to (DN) with the explanandum statement 'Wilbur has the color he has'; for this is a fairly weak statement likely to be implied by some singular premiss of the explanans involving Wilbur.[25] At any rate, the whole difficulty can be avoided more simply, if one so wishes, by excluding predicative expressions as referring expressions, and those of us who are nominalistically inclined would have preferred this course independently of the particular difficulty under discussion.

Let me conclude with a few words concerning the question whether (I) and (L), when suitably revised, should be taken as jointly constituting a necessary condition on the identity of events described by statements. One argument against taking them as such is that it would make all of the following describe or specify distinct events: 'Brutus killed Caesar', 'Brutus assassinated Caesar', 'Brutus stabbed Caesar', and so on. But this argume nt is not conclusive since we are interested in events primarily insofar as they are objects of explanation and relata of causal relation, and it is by no means false or absurd to say that to explain why Brutus stabbed Caesar is not the same as explaining why Brutus assassinated Caesar; and a specification of the causal conditions for the event given by 'Brutus stabbed Caesar' need not be the same as one for the event given by 'Brutus assassinated Caesar'.[26]

The University of Michigan

REFERENCES

[1] Rudolf Carnap, *Philosophical Foundations of Physics*, Basic Books, New York, 1966, p. 193. My italics.

[2] A. I. Melden, *Free Action*, Routledge & Kegan Paul, London, 1961, p. 105. My italics.

[3] Ernest Nagel, *The Structure of Science*, Harcourt, Brace & World, New York, 1961, p. 21. My italics.
[4] A notable recent exception is Donald Davidson's 'Causal Relations', *The Journal of Philosophy* **64** (1967), 691–703. Another interesting recent contribution, which focuses on somewhat different problems from those that will concern us here, is Zeno Vendler, *Linguistics in Philosophy*, Cornell Univ. Press, Ithaca, New York, 1967, chs. 5, 6. Also see R. M. Martin, 'Facts: What They Are and What They Are Not', *American Philosophical Quarterly* **4** (1967), 269–280.
[5] E.g., G. E. M. Anscombe, *Intention*, Basil Blackwell, Oxford, 1957.
[6] The classic statement of this model is contained in Carl G. Hempel and Paul Oppenheim, 'Studies in the Logic of Explanation', reprinted in Hempel, *Aspects of Scientific Explanation*, The Free Press, New York, 1965. For a more recent authoritative study by Hempel see the title essay in the book just cited.
[7] This simple formulation is known to be too broad but we shall use it in this paper, since it at least states an important necessary condition for deductive explanation and there is no known formal characterization of the D-N model that is completely adequate. See Hempel, *ibid.*, pp. 270–278, pp. 291–295, and the further references given therein.
[8] Cf. Arthur Pap: 'In order to strictly deduce the "explanandum" (*fact* to be explained) from the "explanans" (explaining *premises*) we need a universal premise...', *An Introduction to the Philosophy of Science*, The Free Press, New York, 1962, p. 345. My italics. Hereafter we shall often omit 'potential deductive' for the sake of brevity.
[9] Hempel, *op. cit.*, p. 247. Also, cf. Israel Scheffler: 'Let a and b be distinct events, described by the sentences A and B respectively ....' 'Explanation, Prediction, and Abstraction', reprinted in A. Danto and S. Morgenbesser (eds.), *Philosophy of Science*, The World Publishing Company, Cleveland, 1960, p. 275.
[10] I do not wish to attribute this formulation to any particular author, although I believe it can be read into the writings of some philosophers.
[11] It suffices to show that if E and E' are two singular statements which are not logically equivalent, there is an ordered pair (L, C) which is a potential deductive explanans for one of E and E' but not for the other. Assume that E does not logically imply E'. Let $b_1, \dots b_m$ be all the distinct individual constants occurring in E and E'. Then $(x_1) \dots (x_m) [E^* \supset E'^*]$ is the law-like statement obtained by universally generalizing the conditional $[E \supset E']$ with respect to the constants $b_1, \dots b_m$, where E^* and E'^* are like E and E' except for containing an occurrence of x_i where E and E' contain an occurrence of b_i. It is clear, then, that the ordered pair $((x_1) \dots (x_m) [E^* \supset E'^*], E)$ is a potential deductive explanans for E'. On the other hand, the ordered pair is not a potential deductive explanans for E, since E is a deductive consequence of the singular premiss of the ordered pair, namely E itself.
[12] In analogy with 'D-N explanation' we may refer to this or any other similar analysis as a 'D-N analysis of causation'. We ignore here such obvious shortcomings of (3) as the absence of any specification of a temporal order between cause and effect.
[13] Cf. Karl Popper: 'To give a *causal explanation* of an event means to deduce a statement which describes it, using as premises of the deduction one or more *universal laws*, together with certain singular statements, the initial conditions'. *The Logic of Scientific Discovery*, Hutchinson, London, 1959, p. 59. [Translation of *Logik der Forschung* (Springer, Vienna, 1935)].
[14] We ignore here the obvious difficulty with this formulation that causal connections

would disappear from the world if there were no language or no men so that there were no statements.

[15] Although I shall not discuss them here, most of what I shall say about event-describing singular statements applies, *mutatis mutandis*, to nominalized sentences by means of which we often refer to events (or facts), e.g., 'Brutus' stabbing Caesar in 44 B.C.', 'the sinking of the Titanic in 1912'. With regard to these expressions we might ask, for example, whether 'the sinking of the Titanic' and 'the sinking of the fastest ocean liner of 1912' refer to the same event. A natural approach here would be to reduce this question to a question about the statements 'The Titanic sank' and 'The fastest ocean liner of 1912 sank'. On the use of norminalized sentences to refer to events and facts, see the illuminating discussion in Vendler, *op. cit.*

[16] The discussion to follow is largely unaffected by whether or not this extension of the leading idea to predicative expressions is adopted. For example, the crucial trivialization argument from Frege and Quine to be given below is independent of the question of what, if any, predicative expressions designate.

[17] Something like this principle and the ideas leading to it are found in Norman Coles, 'Facts', *Ratio* 7 (1965), 180–189.

[18] See, e.g., W. V. Quine, *The Ways of Paradox*, Random House, New York, 1966, pp. 161–162.

[19] This argument is also used by Davidson for an essentially similar purpose in the paper cited in Reference 4, and also in his 'The Logical Form of Action Sentences', in *The Logic of Decision and Action* (ed. by N. Rescher), University of Pittsburgh Press, Pittsburgh, 1967. As Davidson points out, the argument goes back to Frege and is intimately connected with his theory of reference for sentences.

[20] I am indebted to Fred Feldman for this observation. The problem of locations of events is a complex one; for one possible treatment, see J. Kim, 'On the Psycho-Physical Identity Theory', *American Philosophical Quarterly* **3** (1966), 227–235, especially pp. 231–234.

[21] If it is thought that this holds only if we understand 'Wilbur married Edith' and 'Edith married Wilbur' to mean 'Wilbur intentionally married Edith' and 'Edith intentionally married Wilbur', and that the two statements would then no longer be logically equivalent, we can easily modify our example, thus: 'Wilbur intentionally married Edith who intentionally married him' and 'Edith intentionally married Wilbur who intentionally married her'.

[22] This example can easily be generalized: let F be any property and let b be any object. By 'the F^*-property of b' we mean the property F in case b has F and the property non-F otherwise. Suppose b has F; then $F =$ the F^*-property of b, and hence, by (I), 'b has F' and 'b has the F^*-property of b' describe the same event. But notice what the second statement says: it says just that the object b has F, or non-F, as the case may be.

[23] David Wiggins, 'Identity-Statements' in *Analytical Philosophy* (ed. by R. J. Butler), Basil Blackwell, Oxford, 1965, Second Series. See especially pp. 42–46.

[24] Bromberger speaks of explaining 'why Hector behaves (or behaved) in the way he does (or did)' in his 'An Approach to Explanation' in R. J. Butler, *ibid.*, p. 98.

[25] If we construe 'color' broadly enough so that being colorless counts as a category of being colored, the statement would be implied by any singular premiss implying the existence of Wilbur.

[26] For an excellent defense somewhat along these lines of an essentially similar principle of individuation for actions, see Alvin I. Goldman's forthcoming book, *A Theory of Human Action*, chap. 1.

DONALD DAVIDSON

THE INDIVIDUATION OF EVENTS*

When are events identical, when distinct? What criteria are there for deciding one way or the other in particular cases?

There is a familiar embarrassment in asking identity questions of this sort that comes out clearly if we rephrase the question slightly: when are two events identical? or, when is one event identical with another? It seems only one answer is possible: no *two* events are identical, no event is ever identical with *another*. It is hopeless to try to improve matters by asking instead, when is an event identical with itself? For again, only one answer is possible: always.

The difficulty obviously has nothing special to do with events, it arises in relation to all identity questions. The only move I know for circumventing this conundrum is to substitute for questions about identities questions about sentences about identities. Then instead of asking when events are identical, we may ask when sentences of the form '$a=b$' are true, where we suppose 'a' and 'b' supplanted by singular terms referring to events.

We have no sooner to restate our problem in this standard way, however, than to realize something scandalous about events. Events, even in the best philosophical circles, lead a double life. On the one hand, we talk confidently of sentences that 'describe' or 'refer to' events, and of cases where two sentences refer to the same event; we have grown used to speaking of actions (presumably a species of event) 'under a description'. We characterize causal laws as asserting that every event of one sort is followed by an event of another sort, and it is said that explanation in history and science is often of particular events, though perhaps only as described in one way rather than another. But – and this is the other hand – when we turn to the sentences, formalized in standard ways or in our native dialect, that are so familiarly interpreted as describing or referring to events, or as making universal claims about events, we generally find nothing commonly counted as singular terms that could be taken to refer to events. We are told, for example, that on occasion 'He raised

N. Rescher et al. (eds.), Essays in Honor of Carl G. Hempel.

his arm' and 'He signaled' describe the same action; yet where are the singular terms in these sentences that could do the describing? 'Whenever a piece of metal is heated it expands' is normally taken as quantifying over physical objects and perhaps times; how could we analyze it so as to justify the claim that it literally speaks of events?

Quine has quipped: "No entity without identity" in support of the Fregean thesis that we ought not to countenance entities unless we are prepared to make sense of sentences affirming and denying identity of such entities. But then more obvious still is the motto: 'No identity without an entity', and its linguistic counterpart: 'No statements of identity without singular terms'.

Our problem was to determine when sentences formed by flanking an identity sign with singular terms referring to events are true; at this point the problem seems to invite the response that there are no such sentences because there are no such singular terms. But of course this is too strong; there are singular terms that apparently name events: 'Sally's third birthday party', 'the eruption of Vesuvius in 1906 A.D.', 'my eating breakfast this morning', 'the first performance of *Lulu* in Chicago'. Still, the existence of these singular terms is of uncertain relevance until we can firmly connect such singular terms with sentences like 'Vesuvius erupted in 1906 A.D.' or 'I ate breakfast this morning', for most of our interest in identity sentences about events depends upon the assumption that the singular terms that appear in them refer to entities that are needed for the analysis of more ordinary sentences. If the only pressure for adopting an ontology of events comes from such phrases as 'Sally's third birthday party', we would probably do better to try and paraphrase these away in context than meddle with the logical form of sentences like 'Brutus killed Caesar' or 'Bread nourishes' so as to show singular terms referring to events or variables ranging over them.

Are there good reasons for taking events seriously as entities? There are indeed. First, it is hard to imagine a satisfactory theory of action if we cannot talk literally of the same action under different descriptions. Jones managed to apologize by saying, 'I apologize'; but only because, under the circumstances, saying 'I apologize' *was* apologizing. Cedric intentionally burned the scrap of paper; this serves to excuse his burning a valuable document only because he did not know the scrap was the document and because his burning the scrap was (identical with) his burning the document. Explanation, as already hinted, also seems to call for events. Last week there was a catastrophe in the village. In the course of explaining why it happened, we need to redescribe it, perhaps as an avalanche. There are rough statistical laws about avalanches: avalanches tend to occur when a heavy snow falls after a period of melting and freezing, so that the new snow does not bind to the old. But we could go further in explaining this avalanche – why it came just when it did, why it covered the area it did, and so forth – if we described it in still a different and more precise vocabulary. And when we mention, in one way or another, the cause of the avalanche, we apparently claim that though we may not know such a description or such a law, there must *be* descriptions of cause and avalanche such that those descriptions instantiate a true causal law. All this talk of descriptions and redescriptions makes sense, it would seem, only on the assumption that there are *bona fide* entities to be described and redescribed. A further need for events springs from the fact that the most perspicuous forms of the identity theory of mind require that we identify mental events with certain physiological events; if such theories or their denials are intelligible, events must be individuals. And for such theories to be interesting, there must be ways of telling when statements of event-identity are true.[1]

The reasons just canvassed for accepting an explicit ontology of events rest upon the assumption that one or another currently accepted or debated philosophical position or doctrine is intelligible when taken at face value; so it remains possible to resist the conclusion by rejecting the relevant doctrines as unintelligible, or by attempting to reinterpret them without appeal to events. The prospects for successful resistance are, in my opinion, dim: I do not believe we can give a cogent account of action, of explanation, of causality, or of the relation between the mental and the physical, unless we accept events as individuals. Still, each of these claims needs a detailed defense that I cannot supply here.[2]

There remains, however, a more direct consideration (of which the others are symptoms) in favor of an ontology of events, which is that without events it does not seem possible to give a natural and acceptable account of the logical form of certain sentences of the most common sorts; it does not seem possible, that is, to show how the meanings of such sentences depend upon their composition. The situation may be sketched as follows. It is clear that the sentence 'Sebastian strolled through the streets of Bologna at 2 a.m.' entails 'Sebastian strolled through the streets of Bologna', and does so by virtue of its logical form. This requires, it would seem, that the patent syntactical fact that the entailed sentence is contained in the entailing sentence be reflected in the logical form we assign to each sentence. Yet the usual way of formalizing these sentences does not show any such feature: it directs us to consider the first sentence as containing an irreducibly three-place predicate '*x* strolled through *y* at *t*' while the second contains the unrelated predicate '*x* strolled through *y*'. It is sometimes proposed that we can mend matters by treating 'Sebastian strolled through the streets of Bologna' as elliptical for 'There exists a time *t* such that Sebastian strolled through the streets of Bologna at *t*'. This suggestion contains the seed of a general solution, however, only if we can form a clear idea of how many places predicates of action or change involve. But it is unlikely that we can do this since there appear to be ways of adding indefinitely to the number of places that would be required. Consider, for example, 'The shark devoured Danny by chewing up his left foot, then his left ankle, then his left knee, then ...', or 'The fall of the first domino caused the fall of the last by causing the fall of the second, which caused the fall of the third, which caused....'[3]

Ingenuity may conceive more than one way of coping with these and associated puzzles, but it is impressive how well everything comes out if we accept the obvious idea that there are things like falls, devourings, and strolls for sentences such as these to be about. In short, I propose to legitimize our intuition that events are true particulars by recognizing explicit reference to them, or quantification over them, in much of our ordinary talk. Take as an example 'Sebastian strolled': this may be construed along lines suggested by 'Sebastian took a stroll'. 'There is an *x* such that *x* is a stroll and Sebastian took *x*' is more ornate than necessary, since there is nothing an agent can do with a stroll except take it; thus we may capture all there is with 'There is an *x* such that Sebastian strolled *x*'.

In this way we provide each verb of action or change with an event-place; we may say of such verbs that they take an *event-object*. Adverbial modification is thus seen to be logically on a par with adjectival modification: what adverbial clauses modify is not verbs, but the events that certain verbs introduce. 'Sebastian strolled through the streets of Bologna at 2 a.m.' then has this form: 'There is an event *x* such that Sebastian strolled x, x took place in the streets of Bologna, and x was going on at 2 a.m.' Clearly, the entailments that worried us before go through directly on this analysis.

We recognize that there is no singular term referring to a mosquito in 'There is a mosquito in here' when we realize that the truth of this sentence is not impugned if there are two mosquitos in the room. It would not be appropriate if, noticing that there are two mosquitos in the room, I were to ask the person who says, 'There is a mosquito in the room', 'Which one are you referring to?' On the present analysis, ordinary sentences about events, like 'Doris capsized the canoe yesterday', are related to particular events in just the same way that 'There is a mosquito in here' is related to particular mosquitos. It is no less true that Doris capsized the canoe yesterday if she capsized it a dozen times than if she capsized it once; nor, if she capsized it a dozen times, does it make sense to ask, 'Which time are you referring to?' as if this were needed to *clarify* 'Doris capsized the canoe yesterday'. We learned some time ago, and it is a very important lesson, that phrases like 'a mosquito' are not singular terms, and hence do not refer as names or descriptions do. The temptation to treat a sentence like 'Doris capsized the canoe yesterday' as if it contained a singular term referring to an action comes from other sources, but we should be equally steadfast in resisting it.

Some actions are difficult or unusual to perform more than once in a short or specified time, and this may provide a specious reason in some cases for holding that action sentences refer uniquely to actions. Thus with 'Jones got married last Saturday', 'Doris wrote a check at noon', 'Mary kissed an admirer at the stroke of midnight'. It is merely illegal to get married twice on the same day, merely unusual to write checks simultaneously, and merely good fortune to get to kiss two admirers at once. Similarly, if I say, 'There is an elephant in the bathtub', you are no doubt justified in supposing that one elephant at most is in the bathtub, but you are confused if you think my sentence contains a singular term that refers to a particular elephant if any. A special case arises when we characterize actions in ways that logically entail that at most one action so characterized exists: perhaps you can break a certain piece of news to a particular audience only once; a man can assassinate his enemy only once; a woman can lose her virtue only once. 'Brutus killed Caesar' is then arguably equivalent to 'Brutus killed Caesar exactly once' which is arguably equivalent (by way of Russell's theory of descriptions) to 'The killing of Caesar by Brutus occurred'. This last certainly does contain a description, in the technical sense, of an action, and so we could say that 'Brutus killed Caesar' refers to or describes the killing of Caesar by Brutus in that it is logically equivalent to a sentence that overtly refers to or describes the killing of Caesar by Brutus. By parity of reasoning we should, of course, maintain that 'There exists a prime between 20 and 28' refers to the number 23. There is a good reason against taking this line, however, which is that on this view someone could be uniquely referring without knowing he was using words that imputed singularity.

Confusion over the relation between ordinary sentences about actions, and particular actions, has led some philosophers to suppose or to suggest that these sentences are about *generic* actions, or *kinds* of actions. Von Wright, for example, says that 'Brutus killed Caesar' is about a particular action, while 'Brutus kissed Caesar' is about a generic action.[4] It is true that we can paraphrase 'Brutus kissed Caesar' as 'There is at least one event belonging to the genus, a kissing of Caesar by Brutus'; but we can equally well paraphrase 'Brutus killed Caesar' as 'There is at least one event belonging to the genus, a killing of Caesar by Brutus'. In neither case does the sentence refer to a generic action. Analogous remarks apply to the idea that 'Lying is wrong' is about a kind of action. 'Lying is wrong' may be rendered 'For all x if x is a lie then x is wrong' or even 'The class of lies is included in the class of wrong actions', but neither of these says that a kind of action is wrong, but rather that each action of a kind is wrong.

Failure to find an ordinary singular term referring to an event in a sentence like 'Caesar died' is properly explained by the fact that such sentences are existential and general with respect to events: we do not find a singular term referring to an event because there is none. But many philosophers, not doubting that 'Caesar died' refers to or describes an event, have confusedly concluded that the sentence *as a whole* refers to (or perhaps 'corresponds to') an event. As long ago as 1927, Frank Ramsey pointed out this error, and how to correct it; he described it as the error of conflating facts (which in his view are what sentences or propositions correspond to) and events.[5] And certainly there are difficulties, of a kind more general than we have indicated, with the idea that whole sentences refer to events. For suppose we agree, as I think we must, that the death of Scott is the same event as the death of the author of *Waverley*: then if sentences refer to events, the sentence 'Scott died' must refer to the same event as 'The author of *Waverley* died'. If we allow that substitution of singular terms in this way does not change the event referred to, then a short and persuasive argument will lead to the conclusion that all true sentences refer to the same event. And presumably only true sentences refer to an event; the conclusion may therefore be put: there is exactly one event. Since the argument is essentially the argument used by Frege to show that all sentences alike in truth-value must name the same thing, I spare you the details.[6]

The mistaken view that a sentence like 'Doris capsized the canoe yesterday' refers to a particular event, whether or not tied to the idea that it is the sentence as whole that does the referring, is pretty sure to obliterate the difference between 'Doris capsized the canoe yesterday' and 'Doris' capsizing of the canoe occurred yesterday'. Yet without this distinction firm in our minds I do not believe we can make good sense of questions about the individuation of events and actions, for while the second sentence does indeed contain a singular description (the sentence as a whole meaning 'There is an event identical with the capsizing of the canoe yesterday by Doris'), the first sentence merely asserts the existence of at least one capsizing. If we are not alert to the difference, we are apt to ask wrongheaded questions like: if Jones apologized by saying 'I apologize', do 'Jones apologized' and 'Jones said "I apologize"' describes the same action? The right response is, I have urged, that neither sentence describes an action. We may then add, if we please, that at least one, or perhaps exactly one, action accounts for the truth of both sentences; but both sentences could be true although no apology by Jones was made by his saying, 'I apologize'.[7]

To see how not appreciating the generality in 'Jones apologized' can lead to mistakes about the individuation of events, consider a suggestion of Kim's.[8] Kim assumes that sentences such as 'Brutus killed Caesar' and 'Brutus stabbed Caesar' refer to events, and he asks under what conditions two such sentences describe or refer to the same event. He proposes the following criterion: two sentences are about the same event if they assert truly of the same particulars (i.e., substances) that the same properties (or relations) hold of them. Kim has a rather complicated doctrine of property identity, but it need not delay us since the point to be made depends only on a simple principle to which Kim agrees: properties differ if their extensions do. The effect is to substitute for what I think of as particular, dated events classes of such, and thus to make identities harder to come by. Where I would say the same event may make 'Jones apologized' and 'Jones said "I apologize"' true, Kim is committed to holding that these sentences describe different events. Nor can Kim allow that a stabbing is ever a killing, or the signing of a check the paying of a bill. He must also hold that if psychological predicates have no coextensive physical predicates, then no psychological event is identical with a physical event.

Kim recognizes these consequences of his criterion, and accepts them; but for reasons I find weak. He writes:

> Brutus' killing Caesar and Brutus' stabbing Caesar turn out, on the proposed criterion of event identity, to be different events, and similarly, "Brutus killed Caesar" and "Brutus stabbed Caesar" describe different events. Notice, however, that it is not at all absurd to say that Brutus' killing Caesar is *not the same as* Brutus' stabbing Caesar. Further, to explain Brutus' killing Caesar (why Brutus killed Caesar) is not the same as to explain Brutus' stabbing Caesar (why Brutus stabbed Caesar)....[9]

Certainly Brutus had different reasons for stabbing Caesar than for killing him; we may suppose he went through a little piece of practical reasoning the upshot of which was that stabbing Caesar was a good way to do him in. But this reasoning was futile if, having stabbed Caesar, Brutus has a different action yet to perform (killing him). And explanation, like giving reasons, is geared to sentences or propositions rather than directly to what sentences are about: thus an explanation of why Scott died is not necessarily an explanation of why the author of *Waverley* died. Yet not even Kim wants to say the death of Scott is a different event from the death of the author of *Waverley*. I turn last to Kim's remark that it is not absurd to say that Brutus' killing Caesar is not the same as Brutus' stabbing Caesar. The plausibility in this is due, I think, to the undisputed fact that not all stabbings are killings. We are inclined to say: *this* stabbing might not have resulted in a death, so how can it be identical with the killing? Of course the death is not identical with the stabbing; it occurred later. But neither this nor the fact that some stabbings are not killings shows that this particular stabbing was not a killing. Brutus' stabbing of Caesar did result in Caesar's death; so it was in fact, though of course not necessarily, identical with Brutus' killing of Caesar.

Discussions of explanation may also suffer from confusion about how sentences are related to events. It is sometimes said, for example, that while we explain the occurrence of an event, we can do so only under one or another of its sentential descriptions. In so far as this remark reminds us of the essential intensionality of explanation, it is unexceptionable. But a mistake may lurk. If what we are to explain is why an avalanche fell on the village last week, we need to show that conditions were present adequate to produce *an* avalanche. It would be confused to say we have explained only an aspect of 'the real avalanche' if the reason for saying this lies in the fact that what was to be explained was itself general (for the explanandum contained no mention of a particular avalanche). We might instead have asked for an explanation of why *the* avalanche fell on the village last week. This is, of course, a harder task, for we are now asking not only why there was at least one avalanche, but also why there was not more than one. In a perfectly good sense the second explanation can be said to explain a particular event; the first cannot.

An associated point may be made about causal relations. Suppose it claimed that the lighting of this match was caused by the striking of the match. The inevitable comment (since the time of Mill anyway) is that the striking may have been *part* of the cause, but it was hardly sufficient for the lighting since it was also necessary for the match to be dry, that there be enough oxygen, etc. This comment is, in my opinion, confused. For since this match was dry, and was struck in enough oxygen, etc., the striking of this match was identical with the striking of a dry match in enough oxygen. How can one and the same event both be, and not be, sufficient for the lighting? In fact, it is not *events* that are necessary or sufficient as causes, but events as *described* in one way or another. It is true that we cannot infer, from the fact that the match was struck, and plausible causal laws, that the match lit; we can do better if we start with the fact that a dry match was struck in enough oxygen. It does not follow that more than the particular striking of this match which occurred was required to cause it to light.

Now that we have a clearer idea what it is like to have singular terms, say '*a*' and '*b*', that refer to events we may return to our original question when a sentence of the form '$a=b$' is true. Of course we cannot expect a general method for *telling* when such sentences are true. For suppose

'$(\iota x)(Fx)$' describes some event. Letting 'S' abbreviate any sentence,

$$(\iota x)(Fx) = (\iota x)(Fx \,\&\, S)$$

is true just in case 'S' is true. Since 'S' is an arbitrary sentence, a general method for telling when identity sentences were true would have to include a method for telling when any sentence was true. What we want, rather, is a statement of necessary and sufficient conditions for identity of events, a satisfactory filling for the blank in:

If x and y are events, then $x = y$ if and only if_____.

Samples of answers (true or false) for other sorts of entities are: classes are identical if and only if they have exactly the same members; times are identical if and only if they are overlapped by exactly the same events; places are identical if and only if they are overlapped by exactly the same objects; material objects are identical if and only if they occupy exactly the same places at the same times. Can we do as well as this for events? Here follows a series of remarks that culminate in what I hope is a satisfactory positive answer.

(1) Many events are changes in a substance. If an event a is a change in some substance, then $a = b$ only if b is also a change in the same substance. Indeed, if $a = b$, every substance in which a is a change is identical with a substance in which b is a change. To touch on such necessary conditions of event-identity is to do little more than reflect on what follows if events really do exist; but that is to the present point. And of course we will not alter the event, if any, to which a description refers if in that description we substitute for the name or description of a substance another name or description of the same substance: witness the fact that the death of Scott is identical with the death of the author of *Waverley*. This is an example of a sufficient condition of identity.

We very often describe and identify events in terms of the objects to which they are in one way or another related. But it would be a mistake to suppose that, even for events that are naturally described as changes in an object, we *must* describe them (i.e., produce unique descriptions of them) by referring to the object. For in fact any predicate of any event may provide a unique description: if an event a is F, a may turn out also to be the only event that is F, in which case 'the event that is F' uniquely refers to *a*. One important way to identify events without explicit reference to a substance is by demonstrative reference: 'that shriek', 'that dripping sound', 'the next sonic boom'.

These last points are well made by Strawson.[10] Strawson also remarks that the possibilities for identifying events without reference to objects are limited, because, as he puts it, events do not provide 'a single, comprehensive and continuously usable framework' of reference of the kind provided by physical objects.[11] This claim is made by Strawson in support of a grander thesis, that events are conceptually dependent on objects. According to Strawson we could not have the idea of a birth or a death or a blow without the idea of an animal that is born or dies, or of an agent who strikes the blow.

I do not doubt that Strawson is right in this: most events are understood as changes in a more or less permanent object or substance. It even seems likely to me that the concept of an event depends in every case on the idea of a change in a substance, despite the fact that for some events it is not easy to say what substance it is that undergoes the change.

What does seem doubtful to me is Strawson's contention that while there is a conceptual dependence of the category of events on the category of objects, there is not a symmetrical dependence of the category of objects on the category of events. His principle argument may, I think, be not unfairly stated as follows: in a sentence like 'There is an event that is the birth of this animal' we refer to, or quantify over, events and objects alike. But we can, if we please, express exactly the same idea by saying 'This animal was born' and here there is no reference to, or quantification over, events. We cannot in the same way eliminate the reference to the object.[12] This is supposed to show that objects are more fundamental than events.

A closely related argument of Strawson's is this: the sentence 'The blow which blinded John was struck by Peter' presupposes, for its truth, that John exists, that Peter exists, and that there is a striking of John by Peter. But the last presupposition may also be expressed simply by saying that Peter struck John, which does not treat the blow as an entity on a par with Peter and John. Strawson again concludes that events are dispensable in a sense in which objects are not.[13] It is hard to see how the evidence supports the conclusion.

If 'Peter struck John' and 'There was a striking of John by Peter' express the same presupposition, how can they require different ontologies? If 'This animal was born' and 'There is an event that is the birth of this animal' are genuine paraphrases one of the other, how can one of them be about a birth and the other not? The argument proves either too much or too little. If every context that seems to refer to, or to presuppose, events may be systematically rephrased so as not to refer to events, then this shows we do not need an ontology of events at all. On the other hand if some categories of sentence resist transformation into an eventless idiom, then the fact that we can apparently banish events from other areas cannot suffice to relegate events to a secondary status; indeed it does not even serve to show that the sentences we know how to parse in superficially event-free terms are not about events. It was in fact in just this vein that I have been urging that we cannot give acceptable analyses of 'This animal was born' and 'Peter struck John' without supposing that there are such things as births and blows. In Strawson's view, if I understand him, 'The blow which blinded John was struck by Peter' entails 'Peter struck John'. But a theory about what these sentences mean that justifies the entailment must, or so I have argued, acknowledge an ontology of events. Thus if my interpretation of the evidence is correct, there is no reason to assign second rank to events; while if, contrary to what I have maintained, total reducibility is possible, then again events do not take a back seat, for there are no events.

In my view, a sentence like 'John struck the blow' is about two particulars, John and the blow. The distinction between singular terms and predicates is not abolished: rather, striking is predicated alike of John and of the blow. This symmetry in the treatment of substances and their changes reflects, I think, an underlying symmetry of conceptual dependence. Substances owe their special importance in the enterprise of identification to the fact that they survive through time. But the idea of survival is inseparable from the idea of surviving certain sorts of change – of position, size, shape, color, and so forth. As we might expect, events often play an essential role in identifying a substance. Thus if we track down the author of *Waverley* or the father of Annette, it is by identifying an event, of writing, or of fathering. Neither the category of substance nor the category of change is conceivable apart from the other.[14]

(2) Should we say that events are identical only if they are in the same place? Of course if events have a location, same events have same locations; but here is a puzzle that may seem to cast a doubt on the project of assigning a clear location to events. Perhaps those events are easiest to locate that are obviously changes in some substance: we locate the event by locating the substance. But if one substance is part of another, a change in the first is a change in the second. Every substance is a part of the universe: hence every change is a change in the universe. It seems to follow that all simultaneous events have the same location. The error lies in the assumption that if an event is a change in a substance, the location of the event is the entire space occupied by the substance. Rather, the location of the event at a moment is the location of the smallest part of the substance a change in which is identical with the event.

Does it make sense to assign a location to a mental event such as remembering that one has left a zipper open, deciding to schuss the headwall, or solving an equation? I think we do assign a location to such an event when we identify the person who remembered, decided, or solved: the event took place where the person was. Questions about the location of mental events are generally otiose because in identifying the event we have usually identified the person in whom the event was a change, so no interesting question about the location of the event remains that is not answered by knowing where the person was when the event occurred. When we do not know who the relevant person is, queries about the location of mental events are perfectly in order: 'Where was the infinitesimal calculus invented?'

Mental events (by which I mean events described in the mental vocabulary, whatever exactly that may be) are like many other sorts of events, and like material objects, in that we give their locations with no more accuracy than easy individuation (within the relevant vocabulary) demands. Aside from a few dubious cases, like pains, itches, pricks and twitches, we have no reason to locate mental events more precisely than by identifying a person, for more than this would normally be irrelevant to individuation. Similarly, we uniquely identify a mountain by giving the latitude and longitude of its highest summit, and in one good sense this gives the location of the mountain. But a mountain is a material object, and so occupies more than a point; nevertheless convention decrees no formula for defining its boundaries.

An explosion is an event to which we find no difficulty in assigning a location, although again we may be baffled at a request to describe the total area. The following quotation from an article on locating earthquakes and underground explosions illustrates how smoothly we operate with the concept of the place of an event:

> Information on the accuracy with which a seismic event can be located is not as complete as could be wished.... If data from stations distant from the event are used, it seems realistic to estimate that the site can be located within a circular area whose radius is about eight kilometers. Stations that are 500–2,000 kilometers from the event may give much larger errors....[15]

(3) No principle for the individuation of events is clearer or more certain than this, that if events are identical, they consume identical stretches of time. Yet even this principle seems to lead to a paradox.

Suppose I pour poison in the water tank of a space ship while it stands on earth. My purpose is to kill the space traveller, and I succeed: when he reaches Mars he takes a drink and dies. Two events are easy to distinguish: my pouring of the poison, and the death of the traveller. One precedes the other, and causes it. But where does the event of my killing the traveller come in? The most usual answer is that my killing the traveller is identical with my pouring the poison. In that case, the killing is over when the pouring is. We are driven to the conclusion that I have killed the traveller long before he dies.

The conclusion to which we are driven is, I think, true, so coping with the paradox should take the form of reconciling us to the conclusion. First, we should observe that we may easily know that an event is a pouring of poison without knowing it is a killing, just as we may know that an event is the death of Scott with knowing it is the death of the author of *Waverley*. To describe an event as a killing is to describe it as an event (here an action) that caused a death, and we are not apt to describe an action as one that caused a death until the death occurs; yet it may be such an action before the death occurs. (And as it becomes more certain that a death will result from an action, we feel less paradox in saying 'You have killed him'.[16])

Directness of causal connection may also play a role. To describe the pouring as a killing is to describe it as the causing of a death; such a description loses cogency as the causal relation is attenuated. In general, the longer it takes for the effect to be registered, the more room there is for a slip, which is another way of saying, the less justification there is for calling the action alone the cause.

Finally, there may be a tendency to confuse events described (partly or wholly) in terms of terminal states and events described (partly or wholly) in terms of what they cause. Examples of the first sort are 'the rolling of the stone to the bottom of the hill' (which is not over until the stone is at the bottom of the hill) or 'his painting the barn red' (not over until he has finished painting the barn red); examples of the second sort are 'the destruction of the crops by the flood' (over when the flood is, which may be finished before the crops are) and 'Jones' inviting Smith to the party' (which Jones does only if Smith gets invited, but has finished doing when he drops the card in the mail).[17]

It is a matter of the first importance that we may, and often do, describe actions and events in terms of their causal relations – their causes, their effects, or both. My poisoning of the victim must be an action that results in the victim being poisoned; my killing of the victim must be an action that results in the death of the victim; my murdering of the victim must be an action that results in the death of the victim and also an action that was caused, in part, by my desire for the victim's death. If I see that the cat is on the mat, my seeing must be caused, in part, by the cat's being on the mat. If I contract Favism, I must contract hemolytic anemia as a consequence of eating, or otherwise coming in contact with, the Fava bean. And so forth. This tendency to identify events in terms of their causal relations has deep roots, as I shall suggest in a moment. But it should not lead to a serious difficulty about the dates of events.

(4) Do place and time together uniquely determine an event; that is, is it sufficient as well as necessary, if events are to be identical, that they occupy exactly the same time and the same place? This proposal was made (somewhat tentatively) by John Lemmon;[18] of course the same proposal has often been made for physical objects. I am uncertain both in the case of substances and in the case of events whether or not sameness of time and place is enough to insure identity. Doubt comes easily in the case of events, for it seems natural to say that two different changes can come over the whole of a substance at the same time. For example, if a metal ball becomes warmer during a certain minute, and during the same minute rotates through 35 degrees, must we say these are the same event? It would seem not; but there may be arguments the other way. Thus in the

present instance it might be maintained that the warming of the ball during m is identical with the sum of the motions of the particles that constitute the ball during m; and so is the rotation. In the case of material objects it is perhaps possible to imagine two objects that in fact occupy just the same places at all times but are different because, though never separated, they are separable.

(5) We have not yet found a clearly acceptable criterion for the identity of events. Does one exist? I believe it does, and it is this: events are identical if and only if they have exactly the same causes and effects. Events have a unique position in the framework of causal relations between events in somewhat the way objects have a unique position in the spatial framework of objects. This criterion may seem to have an air of circularity about it, but if there is circularity it certainly is not formal. For the criterion is simply this: where x and y are events,

$$x = y \text{ if and only if } ((\forall z)(z \text{ caused } x \rightarrow z \text{ caused } y) \text{ and } (\forall z)(x \text{ caused } z \rightarrow y \text{ caused } z)).$$

No identities appear on the right of the biconditional.

If this proposal is correct, then it is easy to appreciate why we so often identify or describe events in terms of their causes and effects. Not only are these the features that often interest us about events, but they are features guaranteed to individuate them in the sense not only of telling them apart but also of telling them together. It is one thing for a criterion to be correct, another for it to be useful. But there are certainly important classes of cases at least where the causal criterion appears to be the best we have. If we claim, for example, that someone's having a pain on a specific occasion is identical with a certain complex physiological event, the best evidence for the identity is apt to be whatever evidence we have that the pain had the same causes and the same effects as the physiological change. Sameness of cause and effect seems, in cases like this one, a far more useful criterion than sameness of place and time.[19]

Perhaps sameness of causal relations is the only condition always sufficient to establish sameness of events (sameness of location in space and time may be another). But this should not be taken to mean that the only way of establishing, or supporting, a claim that two events are identical is by giving causal evidence. On the contrary, logic alone, or logic plus physics, or almost anything else, may help do the job, depending on the descriptions provided. What I do want to propose is that the causal nexus provides for events a 'comprehensive and continuously usable framework' for the identification and description of events analogous in many ways to the space-time coordinate system for material objects.

This paper may be viewed as an indirect defense of events as constituting a fundamental ontological category. A defense, because unless we can make sense of assertions and denials of identity we cannot claim to have made sense of the idea that events are particulars. Indirect, because it might be possible to make such needed sense, and to provide clear criteria for identity, and yet to have made no case at all for the need to posit events as an independent category. In other places I have tried to make good on the question of need; here I have not much more than summarized the arguments. But I have found that even those who are impressed with the arguments often have a residual doubt that centers on the apparent intractability of the question when events are identical.

I have tried to banish this doubt as far as I could. The results are not, it must be allowed, overwhelming. But how much should one expect? Can we do any better when it comes to giving criteria for individuating material objects? It should be noticed that the subject has been the individuation of events quite generally, not kinds of events. The analogous problem for material objects would be to ask for conditions of identity of equal generality. At this level, there is individuation without counting. We cannot answer the question, 'How many events occurred (since midnight, between Easter and Christmas)?' but neither can we answer the question, 'How many material objects are there (in the world, in this room)?' We do equally badly on counting classes, points, and intervals of time. Nor are there very good *formulas* for individuating in some of these cases, though we make good enough sense of assertions and denials of identity.

Individuation at its best requires sorts or kinds that give a principle for counting. But here again, events come out well enough: rings of the bell, major wars, eclipses of the moon, and performances of *Lulu* can be counted as easily as pencils, pots, and people. Problems can arise in either domain. The conclusion to be drawn, I think, is that the individuation of events poses no problems worse in principle than the problems posed by the individuation of material objects; and there is as good reason to believe events exist.

Princeton University

REFERENCES

* To my own profit I have discussed questions raised in this paper with P. F. Strawson, David Pears, John Wallace, and David Wiggins. David Kaplan commented on an earlier draft read at a colloquium at the University of California at Irvine in April, 1967, and some of the wisdom in his remarks, if not the wit, has been incorporated in the present draft. My research was supported by the National Science Foundation.

[1] This point is well stated by Jaegwon Kim, 'On the Psycho-Physical Identity Theory', *American Philosophical Quarterly* **3** (1966), 227–235.

[2] For work along these lines, see my 'Actions, Reasons and Causes', *The Journal of Philosophy* **60** (1963), 685–700; 'The Logical Form of Action Sentences', in *The Logic of Decision and Action* (ed. by N. Rescher), Pittsburgh, 1967; my comments on Richard Martin in *Fact and Experience* (ed. by J. Margolis), Oxford, 1969; 'Causal Relations', *The Journal of Philosophy* **64** (1967), 691–703.

[3] The difficulty discussed here is raised by Anthony Kenny in *Action, Emotion and Will,* London, 1964, 2nd ed., chap. VII. In the second and third papers mentioned in the previous reference I devote more space to these matters and to the solution about to be outlined.

[4] Georg Henrik von Wright, *Norm and Action*, London, 1963, p. 23.

[5] 'Facts and Propositions', reprinted in *The Foundations of Mathematics*, New York, 1950, pp. 140, 141.

[6] See 'The Logical Form of Action Sentences', *op. cit.*, pp. 91, 92 and 'Causal Relations', *op. cit.*, pp. 649, 695.

[7] F. I. Dretske in 'Can Events Move?' *Mind* **76** (1967), 479–492, correctly says that sentences do not refer to or describe events, and proposes that the expressions that do refer to events are the ones that can properly fill the blank in 'When did ——— occur (happen, take place)?' This criterion includes (as it should) such phrases as 'the immersion of the paper' and 'the death of Socrates' but also includes (as it should not) 'a discoloration of the fluid'.

[8] In 'On the Psycho-Physical Identity Theory', *op. cit.* Essentially the same suggestion is made by Richard Martin in 'On Events and Event-Description' in *Fact and Experience, op. cit.*

[9] *Op. cit.*, p. 232 (footnote).

[10] *Individuals*, London, 1959, pp. 46ff. I am not sure, however, that Strawson distinguishes clearly between: pointing out an entity to someone; producing a unique description of an entity; producing a description that is guaranteed to be unique.

[11] *Ibid.*, p. 53.

[12] *Ibid.*, pp. 51ff.

[13] *Ibid.*, p. 200.

[14] The same conclusion is reached by J. Moravcsik, 'Strawson and Ontological Priority' in *Analytical Philosophy, Second Series* (ed. by R. J. Butler), Oxford, 1965.

[15] E. C. Bullard, 'The Detection of Underground Explosions', *Scientific American* **215** (1966), 24.

[16] Harry Levin, *The Question of Hamlet*, New York, 1959, p. 35, says in effect that the poisoned Hamlet, in killing the King, avenges, among other murders, his own. This he could not do if he had not already been murdered.

[17] I discuss this issue at greater length in a paper titled 'Agency' to be included in the proceedings of the November, 1968 colloquium on Agent, Action, and Reason at the University of Western Ontario, London, Canada.

[18] E. J. Lemmon, 'Comments on D. Davidson's "The Logical Form of Action Sentences"', in *The Logic of Decision and Action*, *op. cit*. Lemmon goes further, suggesting that '... we may invoke a version of the identity of indiscernables and identify events with *space-time zones*'. But even if there can be only one event that fully occupies a space-time zone, it would be wrong to say a space-time zone *is* a change or a cause (unless we want to alter the language).

[19] Thomas Nagel suggests the same criterion of the identity of events in 'Physicalism', *The Philosophical Review* **74** (1965), 346.

HILARY PUTNAM

ON PROPERTIES

It has been maintained by such philosophers as Quine and Goodman that purely 'extensional' language suffices for all the purposes of properly formalized scientific discourse. Those entities that were traditionally called 'universals' – properties, concepts, forms, etc. – are rejected by these extensionalist philosophers on the ground that 'the principle of individuation is not clear'. It is conceded that science requires that we allow something tantamount to quantification over non-particulars (or, anyway, over things that are not material objects, not space-time points, not physical fields, etc.), but, the extensionalists contend, quantification over *sets* serves the purposes nicely. The 'ontology' of modern science, at least as Quine formalizes it, comprises material objects (or, alternatively, space-time points), sets of material objects, sets of sets of material objects,..., but no *properties*, *concepts*, or *forms*. Let us thus examine the question: Can the principle of individuation for properties ever be made clear?

I. PROPERTIES AND REDUCTION

It seems to me that there are at least two notions of 'property' that have become confused in our minds. There is a very old notion for which the word 'predicate' used to be employed (using 'predicate' as a term only for *expressions* and never for properties is a relatively recent mode of speech: 'Is existence a predicate?' was not a *syntactical* question) and there is the notion for which I shall use the terms 'physical property', 'physical magnitude', 'physical relation', etc., depending on whether the object in question is one-place, a functor, more than one-place, etc. Ignore, if possible, the connotations of 'physical', which are rather misleading (I would be pleased if someone suggested a better terminology for the distinction that I wish to draw), and let me try to tell you what distinction it is that I mean to mark by the use of the terms 'predicate' (which I shall revive, in its classical sense) and 'physical property'.

The principle of individuation for predicates is well known: the prop-

N. Rescher et al. (eds.), Essays in Honor of Carl G. Hempel.

erty of being *P* (where 'property' is understood in the sense of 'predicate') is one and the same property as the property of being *Q* – i.e., to say of something that it is *P* and to say of something else that it is *Q* is to apply the *same predicate* to the two things – just in case '*x* is *P*' is *synonymous* (in the wide sense of 'analytically equivalent to') '*x* is *Q*'. Doubt about the clarity of the principle of individuation for predicates thus reduces to doubt about the notion of synonymy. While I share Quine's doubts about the existence of a clear notion of synonymy, I have more hope than he does that a satisfactory concept can be found, although that is not to be the subject of this paper.

Consider, however, the situation which arises when a scientist asserts that temperature *is* mean molecular kinetic energy. On the face of it, this is a statement of identity of properties. What is being asserted is that the *physical property* of having a particular temperature is *really* (in some sense of 'really') the *same property* as the property of having a certain molecular energy; or (more generally) that the *physical magnitude* temperature is one and the same physical magnitude as mean molecular kinetic energy. If this is right, then, since '*x* has such-and-such a temperature' is not *synonymous* with '*x* has bla-bla mean molecular kinetic energy', even when 'bla-bla' is the value of molecular energy that corresponds to the value 'such-and-such' of the temperature, it must be that what the physicist means by a 'physical magnitude' is something quite other than what philosophers have called a 'predicate' or a 'concept'.

To be specific, the difference is that, whereas synonymy of the expressions '*x* is *P*' and '*x* is *Q*' is required for the predicates *P* and *Q* to be the 'same', it is not required for the physical property *P* to be the same physical property as the physical property *Q*. Physical properties can be 'synthetically identical'.

This fact is closely connected with *reduction*. 'Temperature is mean molecular kinetic energy' is a classical example of a reduction of one physical magnitude to another; and the problem of stating a 'principle of individuation' for physical properties, magnitudes, etc., reduces, as we shall see, to the problem of describing the methodological constraints on reduction. Not all reductions involve properties or magnitudes; for example, 'Water is H_2O' asserts the identity of each body of water with a certain aggregation of H_2O molecules, give or take some impurities, not the identity of 'the property of being water' and 'the property of being

H_2O' – although one might assert that those are the same physical property, too – but many reductions do; e.g., the reduction of gravitation to space-time curvature, of surface tension to molecular attraction, and so on.

I shall suppose, then, that there is a notion of property – for which I use the adjective 'physical', mainly because 'physical magnitude' already exists with a use similar to the use I wish to make of 'physical property', which satisfies the condition that the property *P* can be synthetically identical with the property *Q*, the criterion being that this is said to be true just in case *P* 'reduces' (in the sense of empirical reduction) to *Q*, or *Q* to *P*, or both *P* and *Q* 'reduce' to the same *R*.

II. CAN ONE GET AN EXTENSIONAL CRITERION FOR THE IDENTITY OF PROPERTIES?

The criterion for the identity of properties just given is not extensional, because the relation of reduction is not extensional. Water reduces to H_2O, and H_2O is coextensive with ($H_2O \vee$ Unicorn), but Water does not reduce to ($H_2O \vee$ Unicorn). The difficulty is simply that

$$(x)\,(x \text{ is water} \equiv x \text{ is an aggregation of } H_2O \text{ molecules})$$

is not merely true but nomological ('law-like'), while

$$(x)\,(x \text{ is water} \equiv x \text{ is an aggregation of } H_2O \text{ molecules} \vee x \text{ is a unicorn})$$

is extensionally true (assuming there are no unicorns), but not law-like (unless the non-existence of unicorns is a law of nature, in which case things become still more complicated).

This raises the question: Can one hope to get a criterion for the identity of properties (in the sense of 'physical property') expressible in an extensional language? The problem is related to such problems as the problem of getting a necessary and sufficient condition for 'nomological', and of getting one for causal statements, expressible in an extensional language, and what I shall say about this problem is closely related to the way in which I propose to treat those other problems.

III. FUNDAMENTAL MAGNITUDES

For reasons which will become clear later, I wish to begin by discussing the notion of a fundamental magnitude in physics. It seems clear that no *analytic* necessary and sufficient condition for something to be a fundamental magnitude can be given. At any rate, I shall not even try to give one. But just how serious is this? There do seem to be methodological principles, albeit vague ones, governing the physicist's decision to take certain terms as fundamental-magnitude-terms and not others. Relying on these principles, and on his scientific intuition, the physicist arrives at a list of 'fundamental magnitudes'. At this point he *has* a necessary and sufficient condition for something to be a fundamental magnitude – his list. To be sure, this is an *empirical* necessary and sufficient condition, not an analytic one. But so what? If one has a confirmation procedure, even a vague one, for a term T, and by using that procedure one can arrive at a biconditional of the form $(x)\,(T(x) \equiv \ldots x \ldots)$ that one accepts as empirically true (and the condition $\ldots x \ldots$ occurring on the right side of the biconditional is precise), then what problem of 'explicating the notion of T-hood' remains? Such a term T may be regarded as a *programmatic* term: we introduce it not by a definition, but by a trial-and-error procedure (often an implicit one); and the program is (using the trial-and-error procedure) to find an empirically correct necessary-and-sufficient condition for 'T-hood' which is precise. If this is successful, then the notion of 'T-hood' is precise enough for all scientific purposes. Even if it is unsuccessful, one might succeed in discovering in each individual case whether T applies or not without ever obtaining any general necessary and sufficient condition: if even *this* is unsuccessful, someone is sure to propose that we drop the notion T altogether.

Even if it is not reasonable to ask for an analytic necessary and sufficient condition in the case of programmatic terms, it is surely reasonable to ask for some indication of the associated trial-and-error procedure, provided that we do not demand more precision in the reply than can reasonably be expected in descriptions of the scientific method at the present stage of knowledge. What is the associated 'trial-and-error procedure', or 'confirmation procedure', in the case of the term 'fundamental magnitude'?

One obvious condition is that fundamental magnitude terms must be 'projectible' in the sense of Goodman. Since this is a general requirement on all terms in empirical science, except complex-compound expressions, and since discussing it properly involves (as Goodman rightly stresses), attacking the whole problem of induction, I shall simply take it for granted. (Goodman's solution is, in effect, to say that a term is projectible if we do in fact project it sufficiently often. This leaves the whole problem of why we project some terms to begin with and not others up to psychology. I am inclined to believe that this, far from being a defect in Goodman's approach, is its chief virtue. It is hard to see, once one has passed Goodman's 'intelligence test for philosophers' (as Ullian has described Goodman's discussion of *green* and *grue*), how this question could be anything but a question for psychology. But anyone who feels that there is *some further* philosophical work to be done here is welcome to do it; my feeling is that what we have here is not so much an unsolved philosophical problem as an undefined one.)

A second condition is that these terms must characterize all things – i.e., all particles, in a particle formulation of physics, and all space-time points, in a field formulation of physics. (I believe that one will get different, though interdefinable, lists of fundamental magnitudes depending on which of these two types of formulation one chooses for physics.)

A third condition is that one of these terms must be 'distance', or a term with the aid of which 'distance' is definable, and that the positions of things must be predictible from the values of the fundamental magnitudes at a given time by means of the assumed laws. (This last requirement applies only before 'quantization'.)

A fourth condition is that the laws must assume an especially simple form – say, differential equations (and linear rather than nonlinear, first order rather than second order, etc., as far as possible), if these terms are taken as primitive.

Looking over these conditions, we see that what one has is not one trial-and-error procedure but two. For the laws (or, rather, putative laws) of physics are not fixed in advance, but are to be discovered at the same time as the fundamental magnitudes. If we assume, however, that the laws are to be expressible in a reasonably simple way as differential equations in the fundamental magnitudes, and that statistics enter (in fundamental particle physics) only through the procedure of passing from a deterministic theory to a corresponding quantum-mechanical theory (the so-called procedure of 'quantization'), then the double trial-and-error procedure is reasonably clear. What one does is to simultaneously look for laws expressible in the form specified (which will predict the position of particles), and to look for terms which are 'projectible' and by means of which such laws can be formulated.

To avoid misunderstandings, let me make it clear that I am *not* claiming that it is 'part of the concept' (as people say) of a fundamental law that it *must* be a differential equation, etc. I am saying that that is what we in fact look for *now*. If it turns out that we cannot *get* that, then we will look for the next best thing. We do not know now what the next best thing would be; partly this is a question of psychology, and partly it depends on what mathematical forms for the expressions of laws have actually been thought of at the time. I deny that the double trial-and-error procedure is fixed by *rules* (or, at least, it is a daring and so-far-unsupported guess that it *is* fixed by rules), unless one is willing to count 'look for laws in a form that seems simple and natural' as a *rule*. But the procedure *is* 'fixed' by the *de facto* agreement of scientists on what is a simple and natural form for the formulation of physical laws. It seems to me to be a great mistake in the philosophy of science to overplay the importance of *rules*, and to underestimate the importance of *regularities*. Regularities in what scientists take to be 'simple' and 'natural' may be a matter of psychology rather than methodology; but (a) the line between methodology and psychology is not at all that sharp; and (b) methodology may well *depend* on such psychological regularities.

IV. A CRITERION FOR THE IDENTITY OF 'PHYSICAL$_2$' PROPERTIES

H. Feigl has distinguished two notions of the 'physical'. In Feiglese, every scientific predicate is 'physical$_1$', i.e., 'physical' in the sense of having something to do with causality, space, and time; but only the predicates of *physics* are physical in the narrower sense, 'physical$_2$'. In this terminology, what I have been calling 'physical properties' should have been called 'physical$_1$ properties'. Our problem is to find a criterion for the identity of physical$_1$ properties. In this section I shall approach this problem by discussing the special problem of a criterion of identity for physical$_2$ properties. Assuming that the presently accepted list of fundamental magnitudes is complete, i.e., that there are no further fundamental magnitudes to be discovered, the natural procedure is to correlate physical$_2$ properties with equivalence classes of predicates definable with the aid of the fundamental magnitude terms. Each defined term in the vocabulary of physics (i.e., of elementary particle physics) corresponds to a physical$_2$ property and *vice versa*; two terms correspond to the same physical$_2$ property just in case they belong to the same equivalence class. But what should the equivalence relation be?

There are *two* natural proposals, I think, leading to two quite different notions of physical$_2$ property. One proposal, which I shall not investigate here, would be to take *nomological coextensiveness* as the equivalence relation; the other would be to take *logical equivalence*. I shall choose logical equivalence, because, although we want to allow 'synthetic identities' between physical$_2$ properties and, for example, observation properties (e.g., temperature is mean molecular kinetic energy), it does not seem natural or necessary to consider two terms as corresponding to the same physical$_2$ property when both are already 'reduced' (i.e., expressed in terms of the fundamental magnitude terms), and in their reduced form they are not logically equivalent.

How shall we understand 'logical equivalence', however? I propose to understand 'logical equivalence' as meaning logical equivalence in the narrower sense (not allowing considerations of 'synonymy'); so that P_1 and P_2 will be regarded as corresponding to the same physical$_2$ property only if: (a) P_1 and P_2 are built up out of fundamental magnitude terms alone with the aid of logical and mathematical vocabulary; and (b) $(x)(P_1(x) \equiv P_2(x))$ is a truth of pure logic or mathematics. (The criterion as just given is for one-place predicates; it should be obvious how it is intended to be extended to relations and functors.)

The proposed criterion of identity implicitly takes the stand that *no* relations among the fundamental magnitudes should be considered as 'analytic'. This seems reasonable to me in view of the strongly 'law-cluster' character of the fundamental magnitude terms, but a word of explanation may be in order. Consider, for the sake of an example or two, some of the relations among the fundamental magnitude terms that have seemed analytic in the past. For 'distance' ('$d(x, y)$'), the following relation has often been considered to be 'part of the meaning'; $d(x, y)$ is not equal to zero unless $x = y$. Yet just this relation is given up (for 'space-time distance', at least) by the Minkowskian metric for space-time. Similarly, that $d(x, y)$ has no upper bound is given up when we go over from Euclidean to Riemannian geometry. These examples indicate, to me at any rate, that, when fundamental magnitude terms are involved, it is foolish to regard *any* statement (outside of a logical or mathematical truth) as 'analytic'.

But is it safe to regard even logic and mathematics as analytic? The answer seems to depend on just what is being packed into the notion 'analytic'. If 'analytic' is a covert way of saying 'true by linguistic convention alone', then the view that logic and mathematics are 'analytic' is highly suspect. Certainly I do not presuppose this view (which I do not in any case accept) here. But if 'analytic' means 'true by virtue of linguistic convention *and* logic or mathematics', then trivially all truths of logic or mathematics are 'analytic'. But this thesis is compatible, for example, with the radical thesis that logic and mathematics are empirical, subject to revision for experimental reasons, etc. I do not wish to rule out this attitude towards logic and mathematics (which, in fact, I hold). Thus, when I say that *logical equivalence* is the criterion for the identity of physical_2 properties, I do not mean logical equivalence according to what we today take to be the laws of logic and mathematics; I simply mean equivalence according to whatever may in fact be the truths of logic. If we change our logic, then we may have to change our minds about what physical_2 properties are in fact identical; but the *criterion* of identity will not have changed; it will just be that we made a mistake in its application in some particular cases.

V. BASIC TERMS OF 'NON-FUNDAMENTAL' DISCIPLINES

The issues involved in the reduction of theoretical terms in 'non-fundamental' disciplines to physical_2 terms are so well known by now, that I shall be very brief. (I shall lean on the discussion by Kemeny and Oppenheim, which I regard as still being the best paper on the subject,[1] and on the subsequent paper by Oppenheim and myself).[2] The basic requirement in every reduction, as enunciated by Kemeny and Oppenheim, is that all the observable phenomena explainable by means of the reduced theory should be explainable by means of the reducing theory. This means that the observation terms must be counted as part of the reducing theory –

in the present case, physics – and that we must suppose that we have at least one true biconditional of the form $(x)(O(x) \equiv P(x))$, where P is a physical$_2$ term, for each undefined observation term O. (This requirement is not made by Kemeny and Oppenheim, but it seems the simplest way of ensuring that the maximum possible observational consequences will be derivable from the reducing theory.)

In the paper by Oppenheim and Putnam mentioned above, it is stressed that the reduction at issue need not be made *directly* to physics; if, for example, the laws of psychology are ever reduced to those of cell-biology (explanation in terms of reverberating circuits of neurons, etc.), while the laws of biology are reduced to laws of physics and chemistry, which is itself reduced to physics, then *the laws of psychology will have been reduced to those of physics* from the point of view of the logician of science, even if no one should ever care to write out the definition of a single psychological term directly in physical$_2$ language.

Once one has found a way of explaining the phenomena in physical$_2$ terms (in the sense just explained), then the next step is to see if anything can be found (from the standpoint of the new explanation of the phenomena directly by means of the laws of physics) which answers to the old theoretical primitives. It is not necessary for this purpose that the old laws should be *exactly* derivable from the proposed identification. If we can find a relative interpretation of the old theories into the theory consisting of the laws of physics plus the 'bridge laws' connecting physical$_2$ terms with observation terms, which permits the deduction of a good approximation theory to the old theories, then we identify the things and properties referred to by the basic terms of the old theories with the things and properties referred to by the corresponding physical$_2$ terms (even if some conventional extension of meaning is involved, as in the case of 'water' and 'hot').

On the other hand, it may happen that some basic term of the old theories does not answer to *anything* (in the light of our new way of explaining the phenomena). In this case, we simply *drop* the old theories (or those laws involving the term in question, at any rate) and explain the phenomena by means of 'lower level' theories, including, in the last resort, direct explanation by means of physics-plus-bridge-laws.

The second case is classified by Kemeny and Oppenheim as *reduction by replacement*; the first case is classified as *reduction by means of bi-*

conditionals. Both types of reduction are exemplified in science, and in some cases it is arguable whether, in view of the shifts of meaning involved, a given reduction should be classified as a reduction by replacement or by means of biconditionals. The important point is that *after* the reduction of a discipline, those basic terms that *remain*, that are still regarded as corresponding to 'physical properties' (in the sense of 'physical$_1$') at all, are reduced by means of biconditionals (or identity-signs, as in the case of 'Temperature *is* mean molecular kinetic energy'). For terms which are 'reduced by replacement' are *dropped*, so that the only basic terms that survive are the ones that we reduce by the other method.

VI. PSYCHOLOGICAL PROPERTIES

What first led me to write a paper on the topic of 'properties' was the desire to study reduction in the case of *psychology*. I am inclined to hold the view that psychological properties would be reduced not to physical$_2$ properties in the usual sense (i.e., first-order combinations of fundamental magnitudes), but to *functional states*, where crude examples of the kinds of properties I call 'functional states' would be (a) the property of being a finite automaton with a certain machine table; and (b) the property of being a finite automaton with a certain machine table *and* being in the state described in a certain way in the table. To say that a finite automaton has a certain machine table is to say that *there are properties* (in the sense of physical$_1$ properties) which the object has (i.e., it always has one of them), and which succeed each other in accordance with a certain rule. Thus the property of having a certain machine table is a *property of having properties which…* – although a property of the first level (a property of things), it is of 'second order' in the old Russell-Whitehead sense, in that its definition involves a quantification over (first-order) physical$_1$ properties. This is a general characteristic of all 'functional' properties, as I use the term: although physical$_1$ properties in a wide sense, they are *second order* physical$_1$ properties. How then should a reduction to such properties be analyzed – e.g., *pain* to a certain functional state (as I proposed in an earlier paper)?

The answer is, that if we are willing to accept the hypothesis that all *first-order* physical$_1$ properties will turn out to be reducible to physical$_2$ properties, then all second-order physical$_1$ properties will automatically reduce to *second-order* $physical_2$ properties. If we succeed in reducing psychological properties to properties of the form: *the property (of second-order) of having (first-order) $physical_1$ properties which...*, then we make the further reduction to (second-order) $physical_2$ properties by simply making the theoretical identification of the foregoing $physical_1$ property with the corresponding $physical_2$ property, that is, with *the (second-order) $physical_2$ property of having (first-order) $physical_2$ properties which....*

It is likely, however, that this unusual type of reduction will have to be combined with the more familiar type if psychology is ever to be reduced. For, although a reduction of psychological states to properties of the kind just described would enable us to predict many of the aspects of the behavior of the corresponding species and to understand the functional organization of that behavior, there are undoubtedly aspects of human behavior whose explanation will require a reference not just to the functional organization of the human brain and nervous system, but to the details of the physical realization of that functional organization. An analogous complication has already appeared in the case of the reduction of chemistry to physics, and is beginning to appear in the case of molecular biology. Although many chemical phenomena can be explained 'at the chemical level', in some cases it is necessary to descend to the level of elementary particle physics, even to explain such familiar facts as the liquidity of water and the hardness of diamond; and although many cellular phenomena can be explained at the level of whole cells, nuclei, etc., in the most important cases it is necessary to 'descend' to explanation directly in physical-chemical terms.

It should be noted that if we accept the strict extensionalism which is urged by Quine, then all questions of reduction of properties trivialize upon the passing-over to corresponding questions about *sets*. *Temperature* as a physical magnitude which is not intrinsically quantified has no place in Quine's scheme: instead, we are urged to take as primitive 'temperature-in-degrees-centigrade', or some such. And the statement that temperature *is* mean molecular kinetic energy passes over into the harmless statement that 'temperature in degrees centigrade is directly proportional to mean molecular kinetic energy in c.g.s. units'. I have discussed this difficulty with Quine, and he has suggested meeting it by saying that 'temperature in degrees centigrade *is a quantification of* mean molecular kinetic energy'. (This would indicate why the question '*Why* is temperature in degrees centigrade directly proportional to mean molecular kinetic energy in c.g.s. units?' is not a happy question). Discussing this move would involve discussing: (a) whether it is really satisfactory to think of mean molecular kinetic energy as a class of equivalence-classes as Quine also suggests; and (b) whether the relation 'the function *f* is a quantification of *S*' does not, on the natural understanding of such phrases as 'a quantification of kinetic energy', turn out to be an *intensional* one. Of course, one *can* take the relation extensionally as meaning that temperature is a one-one function of the equivalence-classes, subject to a continuity condition; but then one will not have distinguished between the cases in which one magnitude is a *function* of another, and the cases in which one magnitude *reduces* to another, which is just our problem.

In the same way, there would be no sense, if Quine were right, in discussing whether pain is a brain-state, or a functional state, or yet another kind of state. 'Pain' is construed by Quine as a predicate whose arguments are an organism and a time; if the set of ordered pairs (O, t) such that O is in pain at t is identical with the set of ordered pairs (O, t) such that O satisfies some other condition at t, then *pain* (the relation) *is* (extensionally) the relation that holds between an organism and a time just in case the organism satisfies that other condition. Pain could be *both* a brain-state and a functional state. In some world, pain could even be 'identical' with pricking one's finger – if the organisms in that world experienced pain when and only when they pricked a finger.

Quine does not find this result counterintuitive, because he does not find intensional differences 'philosophically explanatory'. I believe that pointing to differences that are there *is* philosophically explanatory; and it seems to me that these particular differences are 'there'. But I do not expect that either of us will succeed in convincing the other.

VII. PROSPECTS FOR AN EXTENSIONAL CRITERION FOR THE IDENTITY OF PROPERTIES

In the light of the foregoing discussion, I can give a brief answer to the question: can we get a criterion for the identity of properties (in the sense of physical$_1$ properties) which is expressible in extensional language? The answer is that we cannot *today*, as far as I am aware, but that prospects seem reasonably good for obtaining one eventually. The reduction of those observation terms that one might want to take as undefined terms in a reasonable formalization of science seems fully possible, notwithstanding some formidable complexities still to be unravelled. Also, it is assumed in present science that the number of fundamental magnitudes *is* finite (since there are assumed to be only four fundamental kinds of forces); and the assumption that the basic terms of the 'non-fundamental' disciplines will eventually be reduced is at least reasonable.

Of course, the present discussion is entirely empirical in spirit. Indeed, my whole purpose is to break away from two recent traditions ('recent tradition' is deliberate) which seem to me to be sterile and already exhausted: the tradition of 'explication' and the tradition of 'ordinary language analysis'. It may turn out that the number of fundamental magnitudes is infinite; or that some properties other than the ones studied in physics have also to be taken as 'fundamental' (although it is easy to see how the discussion should be modified in this case); or that there are *no* fundamental properties (e.g., there is a level still more 'fundamental' than the level of elementary particles, and a level still more 'fundamental' than that, etc.). If any one of these possibilities turns out to be real, then I am content to leave it to some philosopher of that future to reopen this discussion! The philosophical point that I wish to make is that at present, when we do *not* have a criterion for the identity of arbitrary physical$_1$ properties that is expressible in extensional language, we are still not all that badly off. We do have a criterion for the identity of physical$_2$ properties, as we presently conceive physical$_2$ properties, and this criterion can be extended to other physical$_1$ properties just as rapidly as we succeed in reducing the disciplines in which the corresponding physical$_1$ terms appear to physics. It does not appear unreasonable that we should be unable, in the case of physical$_1$ properties which have not been reduced, to answer the question of identity or non-identity with any certainty prior to the reduction. Of course, in some cases we *can* answer it; for example, properties which are not coextensive are certainly not identical.

VIII. ARE PROPERTIES DISPENSABLE?

That there are many assertions that scientists make that we do not know how to render in a formalized notation without something tantamount to quantification over properties is easily seen. First, consider the question we have mentioned several times: whether there are any fundamental magnitudes not yet discovered. Second, consider the scientist who utters a conjecture of the form 'I think that there is a single property, not yet discovered, which is responsible for such-and-such'. Thirdly, consider the assertion that two things have an unspecified observable property in common.

I believe that all of these cases really reduce to the second: the case of saying that something's having a property P is *responsible for* (or 'causes', etc.) such-and-such. Let us call a description of the form 'the property P, the presence of which (in such-and-such cases) is responsible for (or causes, etc.) such-and-such', a *causal description* of a property. Let us call a description of the form 'the property of being P' a *canonical description* of a property. Then the difficulty is basically this: that there are properties for which we know a causal description but no canonical description. And when we wish to speak of such properties, an existential quantifier over all properties seems unavoidable.

Consider the first case; the case of saying that there is a fundamental magnitude not yet discovered. This comes down to saying that there are phenomena (which itself involves a quantifier over observable properties!) for which some property P is responsible, such that the property P is not definable (in some specified way) in terms of the properties currently listed as 'fundamental'. Consider the third case: quantifying over observable properties. This might be handled in the case of humans by giving a list of all observable properties (although the impracticality of this proposal is obvious); but we also talk of properties that other species can observe and we cannot. But presumably this comes down to talking of those properties P that act as the stimuli for certain responses, and this could presumably be construed as a reference to the properties satisfying certain causal descriptions. Probably, then, (although I do not feel absolutely sure of this) it is ultimately only in causal contexts that quantification over properties is indispensable.

One proposal which has been made for handling such references in an extensional way is this: the assertion that 'A's having the property P at t_0 is the cause of B's having the property Q at t_1,' for example, is handled by saying that '$P(A, t_0)$' is part of an *explanans*, whose corresponding explanandum is '$Q(B, t_1)$'. The 'explanans' and the 'explanandum' are respectively the premiss and the conclusion in an argument which meets the conditions for an explanation as set forth in the familiar covering-law model. Does this obviate the need for property-talk?

I do not wish to discuss here the familiar objections to handling causal statements via the covering-law model (e.g., Bromberger's ingenious objection that this model would permit one to say that the period of a pendulum's being so-and-so *caused* the string to have such-and-such a length). But even without going into the adequacy of this model itself, two points need to be made.

First of all, the proposed analysis of causal statements only works when the properties in question are specified by canonical descriptions. When the property hypothesized to be the cause is mentioned only by a causal description – when part of the explanans is that there *exists* a property with certain causal efficacies – then this analysis does not apply. Of course, one could treat such explanations as programmatic: when the scientist says 'I hypothesize that there is a property which is responsible for such-and-such, and which obey some law of the following form', one could 'translate' this by 'I propose to introduce a new primitive *P* into the language of science, and I suggest that some theory containing the term *P* in such-and-such a way will turn out to be confirmed'; but this is clearly inadequate. (The theory might never be confirmed, because, for example, the money to perform the experiments was not forthcoming, and it might still be true that there *was* a property *P* which..., etc.) Or one might propose to substitute 'is true' for 'will turn out to be confirmed'; but then one runs into the difficulty that one is speaking of 'truth' in a language which contains a primitive one cannot translate into one's meta-language (and which has not, indeed, been given any precise meaning). Or one might say that the scientist has not made any *statement* at all; that he has just said, in effect '*Let's* look for a new theory of the following kind...'; but this seems just plain false.

Secondly, the covering-law theory of explanation uses the term 'nomological' ('law-like') which has admittedly never been explicated. What are the prospects for an explication of this term, in comparison with the prospects of the notion 'property'?

The following would be my program for arriving at a more precise description of the class of 'nomological statements': first, I think that we should try to formulate a hypothesis as to the *form* in which the fundamental laws can be written. This is a much weaker requirement than the requirement that we actually find the laws. The same mathematical form – for example, differential equations in the 'fundamental magnitudes' of classical physics – can be used for the laws of both classical physics and relativity physics. If one takes '$d(x, y)$' (the distance from x to y, where x and y are spacial points at one time) as primitive, then, indeed, this is context-dependent in relativity physics (i.e., $d(x, y)$, or even what x's and y's are spacial points *at one time*, is relative to the reference-system), but this is irrelevant to the statement of the *laws* since these are the same in all reference systems. The change in the geometry is just a change in the laws obeyed by $d(x, y)$; but laws are still expressible as differential equations valid at all points of space and time, and involving only the fundamental magnitudes. Conversely, it seems reasonable to say that any physical relation that can be expressed as a differential equation without boundary conditions, valid at *all* points in space and time, and in just the fundamental magnitudes, should count as a law. Once such a form has been found, the true statements of that form are defined to be the 'nomological' statements of physics. Secondly, as soon as one succeeds in reducing the basic terms of some 'non-fundamental' discipline to physics, one can define the concept 'nomological' for that discipline: namely, a statement in the vocabulary of that discipline is nomological if and only if it is equivalent, via the accepted reducing-biconditionals, to a nomological statement of physics.

It should not be supposed from the foregoing that I think that 'law of nature' *means* 'statement which is reducible to (or which itself is) a nomological statement of physics'. Rather, the situation seems to me to be as follows. Each of the scientific disciplines has pretty much its own list of 'fundamental magnitudes' and its own preferred form or forms for the expression of 'laws'. Thus the discussion in the section of this paper headed 'fundamental magnitudes' could be repeated, with suitable modifications, for each of the other disciplines. In each case there seem to be certain magnitudes which are 'dependent', in the sense that it is the business of the discipline (at least *prima facie* – a discipline may, of course, change its mind about what its 'business' is) to predict their time-course, and certain magnitudes which are independent, in the sense that they are introduced in order to help predict the values of the dependent magnitudes. In physics, for example, it was the position of particles that was, above all, dependent. In economics it would be prices and amounts of production. In each case the scientist looks for a set of properties including his 'dependent variables' which are 'projectible', and which will suffice for the statement of 'laws' – i.e., in order for properties to be accepted as the 'fundamental' ones, it must be possible to formulate (hopefully) true general statements in terms of those properties, which have one of the forms recognized as a form for the expression of laws at that stage in the development of the science, and which will predict, at least statistically, the time-course of the dependent variables.

As we mentioned before in the case of physics, it may be that one cannot *get* true general statements which will do what one wants in the form that one originally takes to be the preferred form for the expression of laws. For example, although the laws of relativity theory can be stated as differential equations in the classical 'fundamental magnitudes', just as the laws of classical physics were, the laws of quantum mechanics require additional 'fundamental magnitudes' (e.g., *spin*), and a more complex mathematical form – one has to introduce operators on Hilbert spaces. Similarly, it might be that the form preferred for the expression of laws in, say, economics at a particular time is too restrictive. When this turns out to be the case, one goes to the 'next best' form, where what is 'next best' is determined by the scientists in the field on the basis of their scientific intuition, in the light of the mathematical methods available at the time for their purposes.

The foregoing suggests that one might seek to explicate the notion of a fundamental law for each of the basic scientific disciplines, and then define a 'nomological statement' simply as a statement which is either itself a fundamental law of one of these disciplines, or which follows from the fundamental laws of one or more disciplines. However, this approach seems to overlook something.

What is overlooked is the enormous impact that the successes of reduction are having on our concept of a natural law. Thus, suppose the Weber-Fechner Law is true without exception (which it in fact is not), and that it is in terms of 'fundamental magnitudes' of psychology, and of the right form to be a 'psychological law'. Then, if it were not for the strong tendency in science in the direction of physicalistic reduction, there would be no question but that it is a 'law of nature'. But let us suppose that when we possess reduction-biconditionals for the concepts involved, we find that the equivalent statement of physics is *not* 'necessary' at the level of physics, and that this is so because there is a perfectly possible mutation such that, if it took place, then the 'Weber-Fechner Law' would fail. In that case, it seems to me that we would not conclude that the 'Weber-Fechner Law' was a natural law, albeit one with an unstated scope limitation (although some ordinary language philosophers have urged such a course), but rather that it was not a law at all, but a good approximation to a more complex statement which *is* a law. It seems to me, in short, that a decisive condition for a statement's being law *is* that the 'equivalent' physical_2 statement be a law of physics, although this decisive condition is itself not part of the 'meaning' of the *word* law, but rather a condition erected by science relatively recently. (Actually, things are much more complicated than the foregoing suggests. For the reductive definitions of the basic terms of the 'non-fundamental' disciplines are themselves selected largely on the basis that they *enable us to derive the laws* of those disciplines. On the other hand, once a reduction has been accepted, there is a great reluctance to change it.)

It is on the basis of the considerations just reviewed that I earlier advanced the suggestion that the following be our program for gaining a precise description of the class of nomological statement: to first try to specify a form in which all the fundamental laws of physics (and only laws, though not necessarily only fundamental ones) can be written; and then to characterize the nomological statements as the statements which follow from true statements of that form together with empirical reduction-biconditionals. (The remaining part of this program – finding a suitable characterization of the law of physics – say, that they all be differential equations in the fundamental magnitudes, valid at every point of space and time, at least 'before quantization' – and finding the empirical reduction-biconditionals – is, of course, a task for science and not for philosophy. In a sense, that is the whole point of this paper.)

It is evident that if this particular program for characterizing the nomological statements ever succeeds, so must the program for characterizing 'identity of properties'. Indeed, the program for characterizing the nomological statements is in one way harder of fulfillment than the program for characterizing identity of properties, in that the latter program requires only that we know the reduction-biconditionals (and reduction-identities) and the fundamental magnitudes of physics, but not that we ever have a suitable characterization of the form of the physical laws.

What of the more conventional program for explicating 'nomological', as represented by the writing of Reichenbach, Goodman *et al.*? This program is to characterize fundamental laws (in all disciplines at once) as *true generalizations of a certain form*, where the specification of the form involves not only considerations of logical form in the usual sense, but such restrictions on the choice of the predicates as Goodman's requirement of projectibility. It is, further, a part of this program to be independent of the sorts of empirical considerations that I have constantly been bringing up – one is apparently, to discover a form in which all and only laws of nature can be written (i.e., no *true* statement can have the form in question and not be a law of nature), and to do this by reflection on the meaning of 'law' (and, perhaps, on the methodology of science) alone. In short, what I hope scientists may be able to do empirically in the next couple of hundred years these philosophers would like to do *a priori*. Good luck to them!

It should be noted that if these philosophers ever succeed, then they will also succeed in providing us with one criterion for the identity of properties (though not the one I have suggested): for *nomological equivalence*, in spite of some counterintuitive consequences, is another criterion for the identity of $physical_1$ properties that I think would be workable (if one were willing to change one's intuitions a bit), and that deserves further investigation. Moreover, if one *could* 'explicate "nomological"', then one should also be able to explicate 'reduction law', and hence to explicate the criterion for the identity of $physical_1$ properties suggested in this paper.

In terms of the foregoing discussion, my answer to the question of whether quantification over properties is indispensable goes as follows: First, there are important locutions which are most naturally formalized as quantifications over properties, and for which there is no obvious formalization today in extensional language. Secondly, the concept of a property is intimately connected with the notions: *nomological*, *explanation*, *cause*, etc., and even comes close to being definable in terms of these notions. Yet these notions are generally admitted to be indispensable in science, even by those philosophers who reject *analytic*, *necessary*, *synonymy*, etc., (i.e., the notions most closely connected with the *other* concept of property mentioned at the beginning of this paper, the concept of a predicate, or the concept of a concept). The notion *is* indispensable, then, in the sense in which any notion is (i.e., we might use different *words*, but we would have to somehow be able to get the notion expressed); and, if the discussion of the prospects for a criterion of identity earlier in this paper was not unduly optimistic, science is well on its way to giving us as precise a criterion for the identity of properties as we could ask for. Let us, then, keep our properties, while not in any way despising the useful work performed for us by our classes!

Harvard University

REFERENCES

[1] J. Kemeny and P. Oppenheim, 'On Reduction', *Philosophical Studies* 7 (1956), 6–19.
[2] P. Oppenheim and H. Putnam, 'Unity of Science as a Working Hypothesis', *Minnesota Studies in the Philosophy of Science*, vol. II (ed. by H. Feigl, G. Maxwell, and M. Scriven), University of Minnesota Press, Minnesota, 1958, pp. 3–36.

FREDERIC B. FITCH

A METHOD FOR AVOIDING THE CURRY PARADOX

The Curry paradox[1] may be formulated in the following way:[2] Let 'Y_q' serve as an abbreviation for '$\hat{x}\,[[x \in x] \supset q]$'. Then, using the method of subordinate proofs, we have:

1	$\quad \vert\ Y_q \in Y_q$	hypothesis
2	$\quad \vert\ Y_q \in \hat{x}\,[[x \in x] \supset q]$	1, repetition and definition
3	$\quad \vert\ [Y_q \in Y_q] \supset q$	2, abstraction elimination
4	$\quad \vert\ q$	1, 3, implication elimination
5	$[Y_q \in Y_q] \supset q$	1–4, implication introduction
6	$Y_q \in \hat{x}\,[[x \in x] \supset q]$	5, abstraction introduction
7	$Y_q \in Y_q$	6, repetition and definition
8	q	5, 7, implication elimination

Thus a procedure appears to exist for giving a proof of any arbitrary sentence 'q'. In particular, the sentence thus proved might be a contradictory statement such as '$r \,\&\! \sim r$'. This situation is paradoxical at least in the sense that ordinary implication rules, when combined with unrestricted abstraction (attribute) rules, give contradictory results. In a similar way the well-known Russell paradox shows that ordinary negation rules, when combined with unrestricted abstraction rules, also give contradictory results.[3] It is desirable not only to find intuitively acceptable methods for avoiding the Curry and Russell paradoxes, but also to find intuitively acceptable systems of logic that can be shown to contain no contradictions whatever. In fact, the least that should be required of a system of *logic* is that it should be *demonstrably consistent*. It might not be unreasonable to go even further and demand that every theorem of such a system should be, in some obvious way, *logically true*. Even the weaker demand, however, is not known to be satisfied either by the system of the simple theory of types or by the standard systems of set theory when the axiom of infinity is included in these various systems.

In order to avoid the Curry paradox or the Russell paradox, it would seem to be sufficient to restrict the abstraction rules in some suitable way, for example by invoking the simple theory of types, so that such expressions as '$x \in x$' would be rejected as not well formed. But all proposed restrictions on abstraction, including those of type theory and those of set theory, seem unduly arbitrary and, in the case of the simple theory of types and standard set theory, are not even known to guarantee freedom from contradiction. Consequently there is a good deal of theoretical interest in seeing how far one can go without placing any restriction at all on abstraction. If the abstraction rules are to be used as embodying a valid logical principle, namely the principle of abstraction, then it is hard to see how any limitation can be placed on abstraction that is not itself a purely logical limitation. But there do not seem to be such limitations that are both purely logical and sufficiently non-arbitrary. In other words, our logical intuitions do not appear to tell us clearly exactly how much leeway to grant to the principle of abstraction, so that if we accept it at all as a principle of logic, we should apparently accept it without limitation.

Previously I have proposed two different restrictions that would avoid the Curry paradox without directly limiting the abstraction rules. The first of these is called *the special restriction* and the second is called *the simple restriction.*[4] I never was very happy about either of these restrictions, and now I have found a different restriction that is far more satisfactory. Before presenting the latter, however, it might be useful to review briefly these two earlier restrictions and to describe a restriction suggested by Prawitz.[5]

The special restriction asserts, roughly, that a proof is not acceptable if somewhere in it a sentence is derived from a subproof that has that sentence as an hypothesis and that has at least one other sentence as an item. Thus, in the foregoing formulation of the Curry paradox, the special restriction is violated because the sentence at step 7 is derived from the subproof 1–4 which has this same sentence as its hypothesis. The simple restriction, on the other hand, asserts that there cannot be two items of a proof or subproof such that the earlier item is justified by an introduction rule and the later item by an elimination rule.[6] The pair of items at steps 5 and 8 violate the simple restriction, and so does the pair at steps 6 and 8.

Not only does each of these restrictions provide a way for avoiding the Curry paradox, but either of these restrictions may be employed in connection with the system S of my book *Symbolic Logic*, and will then guarantee the consistency of that system. In particular, the Russell paradox as well as the Curry paradox will be avoided in S if either of these restrictions is used. A disadvantage of the special restriction is that it requires searching through a proof to see if any sentences are derived from subproofs of the wrong kind. A disadvantage of the simple restriction is that it rules out some patently acceptable types of argument such as the following:

1	p	hypothesis
2	q	hypothesis
3	$[p \& q] \supset r$	hypothesis
4	$p \& q$	1, 2, conjunction introduction
5	r	3, 4, implication elimination

A restriction proposed by Prawitz is not subject to the above disadvantages.[7] His statement of this restriction may be paraphrased as follows: No items of a proof or subproof may serve both as direct consequences with respect to an introduction rule and as the major premisses of an elimination rule. (The premisses of all elimination rules will be said to be major premisses, except that in the case of implication elimination we call 'p' the minor premiss and '$p \supset q$' the major premiss.) The above example of an intuitively acceptable argument does not violate Prawitz's restrictions, though it does violate the simple restriction. The Curry paradox, on the other hand, does violate Prawitz's restriction since the item at step 5 is a major premiss of an application of implication elimination while at the same time being a direct consequence by implication introduction. Use of Prawitz's restriction makes possible a proof of the consistency of system S, just as in the case of the other two restrictions.

None of the restrictions so far discussed makes possible a proof of the consistency of S, however, if to S we add a rule of negative implication introduction according to which '$\sim [p \supset q]$' is a direct consequence of 'p' and '$\sim q$'. It is very desirable to be able to add such a rule to S, and also to be able to add the rule of negative implication elimination according to which 'p' and '$\sim q$' are direct consequences of '$\sim [p \supset q]$'. These two rules can be added to S and consistency still proved if the restriction described as follows is employed: A sentence which is an item of a proof or subproof cannot also be an hypothesis of a subproof which is another item of that proof or subproof.[8] This will be called the *restriction of nonrecurrence* since it precludes an item of a proof or subproof from recurring as an hypothesis of an item of that proof or subproof. This restriction is violated by the above formulation of the Curry paradox since the sentence at step 7 occurs both as the hypothesis of an item of the proof and also as an item of the proof.

This new restriction of nonrecurrence is very easy to apply. It is slightly stronger than Prawitz's restriction, something to be expected since it will provide consistency of **S** in the presence of the negative implications rules. It is at least as strong as Prawitz's restriction in the sense that any proof conforming to it can be transformed into an equivalent proof that satisfies Prawitz's restriction. More specifically, consider any proof that conforms to the restriction of nonrecurrence but contains an item that violates Prawitz's restriction. The use of this item as a major premiss of an elimination rule can always be shown to be unnecessary in the sense of contributing nothing new, so that the application of the elimination rule can be dropped, thus allowing the item to conform to Prawitz's restriction after all. For example, suppose that an item of the form '$p \& q$' violates Prawitz's restriction. Then this item must be a direct consequence by an introduction rule and it must also be a major premiss for an elimination rule. But the only introduction rule that gives it as a direct consequence is conjunction introduction (the rule that asserts that '$p \& q$' is a direct consequence of 'p' and 'q'), and the only elimination rule for which it serve as a major premiss is conjunction elimination (the rule that asserts that 'p' and 'q' are direct consequences of '$p \& q$'). Hence the application of conjunction elimination can be dropped, since all that conjunction elimination can give, namely 'p' and 'q', must already be present as premisses from which '$p \& q$' was obtained by conjunction introduction, and the violation of Prawitz's restriction has been removed without genuine loss and even, in this case, without invoking the restriction of nonrecurrence.

Consider, next, a similar case that does, however, require invoking the restriction of nonrecurrence. Suppose that an item of the form '$p \supset q$' violates Prawitz's restriction. Then there must be an item 'q' that is obtained from 'p' and '$p \supset q$' by implication elimination, while the item '$p \supset q$' itself is obtained by implication introduction. We wish to show that the application implication elimination can be dropped without loss and the violation of Prawitz's restriction thereby avoided. This can be done by showing that 'q' must have been present as a previous item, so that nothing is gained in obtaining it by an application of implication elimination. The argument is as follows: Since '$p \supset q$' was obtained by implication introduction, at least one of two alternatives must hold: Either (1) '$p \supset q$' is preceded by a subproof having hypothesis 'p' and conclusion 'q', or else (2) '$p \supset q$' is preceded by 'q' itself. (These are the only ways of obtaining '$p \supset q$' by implication introduction, and it is of course understood that in the first case the preceding subproof is an item of the same proof or subproof as the item '$p \supset q$' itself, while in the second case the preceding sentence 'q' is, similarly, an item of the same proof or subproof as the item '$p \supset q$' itself.) But the first of these two alternatives is ruled out by the restriction of nonrecurrence, since under the first alternative the sentence 'p' would be serving as an item of some proof or subproof and also as an hypothesis of a subproof which is itself an item of that same proof or subproof. There remains, then, only the second alternative which guarantees that 'q' has already appeared as a preceding item. Similar types of argument can be used to show that other kinds of violation of Prawitz's restriction can be avoided in the presence of the restriction of nonrecurrence. The latter restriction need be invoked, in fact, only for the case just considered.

A consequence of the fact that every proof in the system S can be put into a form satisfying Prawitz's restriction is this: Every categorical main proof (that is, every proof that has no hypotheses and is not subordinate to another proof) can be put into a form that uses only introduction rules for its own items (though proofs subordinate to it might use elimination rules). This is because a main proof that has no hypotheses can begin only by use of introduction rules that have no premisses (that is, by use of *axiom rules*, or *axioms*, such as the rule of identity introduction which has no premisses and which permits '$a = a$' to appear axiomatically as an item of any proof or subproof), and thereafter all further items of such a proof have to be obtained by introduction rules on pain of violating Prawitz's restriction.

This fact, that every categorical main proof need use only introduction rules, in turn guarantees the consistency of the system S, since it can be shown that contradictory pairs of sentences are obtainable by introduction rules only from previous pairs of contradictory sentences. For example, if a contradictory pair of sentences of the form '$p \& q$' and '$\sim [p \& q]$' is such that each of them is an item of some proof by applications of introduction rules, then this contradictory pair must have been obtained in this proof either from the contradictory pair 'p' and '$\sim p$', or else from the contradictory pair 'q' and '$\sim q$'. More specifically, 'p' and 'q' must both have appeared as previous items (since otherwise '$p \& q$' could not have been obtained by an introduction rule, the only applicable introduction rule in this case being conjunction introduction), and furthermore either '$\sim p$' or '$\sim q$' must have appeared as a previous item (since otherwise '$\sim [p \& q]$' could not have been obtained by an introduction rule, the only applicable introduction rule in this case being the rule of negative conjunction introduction which asserts that '$\sim [p \& q]$' is a direct consequence of '$\sim p$' and also a direct consequence of '$\sim q$'). Thus either the contradictory pair 'p' and '$\sim p$', or the contradictory pair 'q' and '$\sim q$' must have preceded the contradictory pair '$p \& q$' and '$\sim [p \& q]$'. In general, when introduction rules are being used, a contradictory pair can never appear as items of a categorical proof, because the axioms (that is, introduction rules without premisses) at the beginning do not supply any contradictory pair, and a contradictory pair cannot appear unless some other contradictory pair has previously appeared. The conclusion is that the system **S** is consistent if the restriction of nonrecurrence is used, since no contradictory pair of sentences can be items of a categorical proof.

This argument for the consistency of **S** also holds when **S** has been extended by adding to it the negative implication rules that were mentioned earlier. In fact, a great advantage of the restriction of nonrecurrence is that it permits the addition of these rules to **S** without destroying consistency. In order to see how part of the argument runs when these rules are included, let us suppose that a categorical proof has a contradictory pair of items of the form '$p \supset q$' and '$\sim [p \supset q]$'. Then it can be shown as follows that 'q' and '$\sim q$' must be a preceding pair of contradictory items: The categorical proof may be assumed to use only introduction rules for its own items. Consequently, 'p' and '$\sim q$' must be preceding items, since only from them can '$\sim [p \supset q]$' follow by an introduction rule, in this case the rule of negative implication introduction. Furthermore, either there is a preceding item which is a subproof with hypothesis 'p' and conclusion 'q', or else 'q' itself is a preceding item, since only from such preceding items can '$p \supset q$' follow by an introduction rule, in this case the rule of implication introduction. But the first of these two alternatives is ruled out by the restriction of nonrecurrence, since 'p' itself is an item of the proof. Hence 'q' and '$\sim q$' are preceding items, as was to be shown.

In order to avoid excessive detail, attention has been restricted to the role played by conjunction and implication in the argument for the consistency of system S as guaranteed by the restriction of nonrecurrence. But S also has rules for negation and disjunction, and these rules can be fitted into the total argument in very much the same way as the rules for conjunction and implication.

There is in S no rule of negation introduction. The rule of negation elimination asserts that any sentence 'q' is a direct consequence of contradictory sentences 'p' and '$\sim p$'. The rule of double negation introduction asserts that '$\sim(\sim p)$' is a direct consequence of 'p', and the rule of double negation elimination asserts, conversely, that 'p' is a direct consequence of '$\sim(\sim p)$'. The rule of disjunction introduction asserts that '$p \vee q$' is a direct consequence of 'p' and also a direct consequence of 'q'. The rule of disjunction elimination asserts that 'r' is a direct consequence of '$p \vee q$' and two subproofs, one with hypothesis 'p' and conclusion 'r' and the other with hypothesis 'q' and conclusion 'r'. The rule of negative disjunction introduction asserts that '$\sim [p \vee q]$' is a direct consequence of '$\sim p$' and '$\sim q$', and the rule of negative disjunction elimination asserts, conversely, that '$\sim p$' and '$\sim q$' are direct consequences of '$\sim [p \vee q]$'. There is also a negative conjunction elimination rule, not mentioned so far, which asserts that 'r' is a direct consequence of '$\sim [p \& q]$' and two subproofs, one with hypothesis '$\sim p$' and conclusion 'r' and the other with hypothesis '$\sim q$' and conclusion 'r'. Conjunction could be treated as defined in terms of disjunction and negation, or disjunction could be treated as defined in terms of conjunction and negation. The formulation given above of the negative rules for conjunction and disjunction is different from that given in my *Symbolic Logic*, but is equivalent to it.

A remark should be made about the rule of negation elimination. This rule, as noted above, asserts that any sentence 'q' is a direct consequence of contradictory sentences 'p' and '$\sim p$'. When this elimination rule is included among the other elimination rules, then some of the arguments previously given have to be slightly modified. In particular, for example, it is no longer true that conjunction elimination is the only elimination rule for which '$p \& q$' can serve as a major premiss, or that negative conjunction elimination is the only elimination rule for which '$\sim[p \& q]$' can serve as a major premiss, because these two sentences might together be the two major premisses for an application of negation elimination, and indeed they might both themselves have been obtained by introduction rules even while serving as such premisses, thus violating Prawitz's restriction. (If only *one* of them was obtained by an introduction rule, Prawitz's restriction would not be violated in allowing them to serve as major premisses for negation elimination.) In connection with such a violation of Prawitz's restriction, however, we should recall that contradictory sentences obtained by applications of introduction rules will always be preceded by other contradictory sentences, so that any result obtained by use of negation elimination from such contradictory sentences can likewise be obtained from earlier contradictory sentences and ultimately from some contradictory sentences that are not obtained by introduction rules, since we know that introduction rules by themselves cannot give contradictory results. Thus any such violation of Prawitz's restriction, involving the rule of negation elimination in the way indicated, can be avoided in the presence of the restriction of nonrecurrence.

The abstraction (attribute) rules and negative abstraction rules can easily be fitted into the general argument for the consistency of **S**, and so can the identity rules. The rule of abstraction introduction asserts that '$a \in \hat{x}(\ldots x \ldots)$' is a direct consequence of '$(\ldots a \ldots)$', while the rule of abstraction elimination asserts that the latter is a direct consequence of the former. Similarly the rule of negative abstraction introduction asserts that '$\sim[a \in \hat{x}(\ldots x \ldots)]$' is a direct consequence of '$\sim(\ldots a \ldots)$', and the rule of negative abstraction elimination asserts, conversely, that '$\sim(\ldots a \ldots)$' is a direct consequence of '$\sim[a \in \hat{x}(\ldots x \ldots)]$'. Bound variables may be thought of as ranging over individuals, propositions, and attributes. The rule of identity introduction, as already noted, asserts that '$a = a$' may be an item of any proof or subproof. The rule of negative identity introduction asserts that '$\sim[a = b]$' may be an item of any proof or subproof provided, that 'a' and 'b', when written in unabbreviated form, are different expressions. There are no *underived* elimination rules for identity. There is a *derived* rule of substitution of equals for equals called identity elimination. There is also another derived rule that asserts, in effect, that all sentences of the form '$a=b$' satisfy the principle of excluded middle.

Quantifier rules can be included in the system **S** by providing rules for universal quantification that are analogous to the rules for conjunction, and by providing rules for existence quantification that are analogous to the rules for disjunction.[9] When such rules are added to the system **S**, the resulting system can still be proved consistent in very much the same way as before.

It may be mentioned that even though the system **S** is nonclassical in the respect that excluded middle holds only in some cases, mainly in the case of identities and various combinations of identities, nevertheless, using the theory of implication sketched above, it is possible to define restricted kinds of quantifiers that in effect refer only to things that do satisfy excluded middle. If the things thus referred to, that is, the things that satisfy excluded middle, are regarded as being propositions by definition, then the system **S** not only provides for quantification over propositions (in this sense of *proposition*), but there would be a theorem of **S** asserting that every proposition satisfies the principle of excluded middle. Thus, from this point of view, **S** can be considered to be classical after all. Another consequence of this point of view is that the expression '$Y_q \in Y_q$' of our formulation of the Curry paradox would not be regarded as expressing a proposition at all, since, under the restriction of nonrecurrence, we can prove neither '$Y_q \in Y_q$' nor '$\sim[Y_q \in Y_q]$', and so cannot prove '$[Y_q \in Y_q] \vee \sim[Y_q \in Y_q]$'. On the other hand, '$[Y_q \in Y_q] \supset q$' would be provable and so would be regarded as expressing a proposition. But '$Y_q \in Y_q$' is a direct consequence of '$[Y_q \in Y_q] \supset q$' by the rule of abstraction introduction, even though the former is not provable and the latter is provable. Thus the restriction of nonrecurrence does have the side effect of slightly limiting abstraction. This new restriction, nevertheless, appears to be the simplest and most reasonable one for purposes of achieving a proof of consistency while still leaving intact to the system **S** a rich logical and mathematical content.

The kind of consistency that has so far been considered may be called *weak consistency*, defined as follows: A system will be said to be *weakly consistent* if it contains no theorem of the form '$p \,\&\!\sim\! p$'. On the other hand a system will be said to be *strongly consistent* if it contains no theorem 'p' such that '$\sim p$' is also a theorem of it. The restriction of nonrecurrence, even in the presence of the negative implication rules, guarantees the weak consistency of S. This is because it guarantees that there is no proof having contradictory items and hence no proof having a sentence of the form '$p \& {\sim} p$' as an item. If the negative implication rules are omitted from S, then the restriction of nonrecurrence can be shown to guarantee the strong consistency of S just as does the special restriction under the same circumstances.[10]

If the negative implication rules are included in S, then although either the special restriction or the restriction of nonrecurrence will guarantee the weak consistency of S, neither of these restrictions guarantees the strong consistency of S, and in the case of both these restrictions the system S fails to be strongly consistent. This was shown for the case of the special restriction independently by H. P. Galliher and J. R. Myhill,[11] and for the case of the restriction of nonrecurrence by Alan R. Anderson in a letter to me. The same example can actually be used in the case of both restrictions, though the example that Anderson devised differs from that of Galliher and Myhill. He defined 'Y' as '$\hat{x}[[x \in x] \supset {\sim}[x \in x]]$' and proved both '$[[Y \in Y] \supset {\sim}[Y \in Y]] \supset {\sim}[[Y \in Y] \supset {\sim}[Y \in Y]]$' and its denial.

The types of inconsistency that arise when S is weakly consistent but not strongly consistent are innocuous in the sense that, after all, the basic theorems that one wants for the foundations of mathematics remain provable and their denials remain unprovable.[12] Strong consistency, as already noted, is easily achieved at the expense of omitting the negative implication rules, and such omission actually causes no serious inconvenience in dealing with the foundations of mathematics (as can be seen by examining the system K' referred to above). The present standpoint, accordingly, provides a choice between a strongly consistent basis for mathematics and a weakly consistent basis. The basis for mathematics provided by standard set theory, by contrast, has the defect that we are completely ignorant of its consistency or lack of consistency.

Yale University

REFERENCES

[1] H. B. Curry, 'The Inconsistency of Certain Formal Logics', *The Journal of Symbolic Logic* 7 (1942), 115–117. Also, H. B. Curry and R. Feys, *Combinatory Logic*, North-Holland Publ. Co., Amsterdam, 1958, pp. 258–262.

[2] This formulation of the paradox is essentially the same as that given in my book, *Symbolic Logic*, Ronald Press, New York, 1952, pp. 258–262. The present essay does

not attempt to present all the details of the system **S**, or even to describe it in exactly the way it is described there, but rather this essay outlines some of the leading ideas involved in the proof of the consistency of **S**, especially as these ideas can be developed in connection with the newly proposed restriction of nonrecurrence. There is no discussion of such topics as general subproofs or reiteration. The conventions regarding quotation marks are assumed to be the same, in principle, as those used in most of my papers in *The Journal of Symbolic Logic*.

[3] *Symbolic Logic, op. cit.*, pp. 109–111.

[4] *Symbolic Logic, ibid.*, pp. 106–109.

[5] D. Prawitz, *Natural Deduction,* Almqvist & Wiksell, Stockholm, 1965, pp. 94–97.

[6] In applying this restriction, the assumption would have to be made that the order of items could not be changed by use of the rule of repetition (*Symbolic Logic, op. cit.*, pp. 35–36), a rule which is not essential anyhow.

[7] D. Prawitz, *loc. cit.*

[8] In applying this restriction we should, strictly speaking, first suppose that all reiterations that could be made have been made. For the concept of reiteration see *Symbolic Logic, op. cit.*, pp. 20–21.

[9] When quantification is treated in this way as infinite conjunction or disjunction, the resulting quantifier rules are said to be *nonfinitary*, and the resulting system should perhaps be referred to as **S′** since it is essentially the same as the nonfinitary system **S′** of Section 27 of *Symbolic Logic*. The consistency of such a nonfinitary system implies the consistency of the corresponding finitary systems.

[10] Cf. pp. 125–127 of my *Symbolic Logic, op. cit.*

[11] The footnote on p. 127 of *Symbolic Logic, op. cit.*

[12] For example, such theorems as the theorems of the system K' in my papers in volumes **15** and **16** of *The Journal of Symbolic Logic*.

PUBLICATIONS (1934–1969) BY CARL G. HEMPEL

1. *Beiträge zur logischen Analyse des Wahrscheinlichkeitsbegriffs*, Ph.D. Thesis, Berlin, 1934. 72 pp.
2. 'On the Logical Positivists' Theory of Truth', *Analysis* **2** (1935), 49–59.
3. 'Analyse logique de la psychologie', *Revue de Synthèse* **10** (1935), 27–42. (English translation, under the title 'The Logical Analysis of Psychology', in *Readings in Philosophical Analysis* [ed. by H. Feigl and W. Sellars], Appleton-Century-Crofts, Inc., New York, 1949, pp. 373–384.)
4. 'Zur Frage der wissenschaftlichen Weltperspektive', *Erkenntnis* **5** (1935/36), 162–164.
5. Über den Gehalt von Wahrscheinlichtkeitsaussagen', *Erkenntnis* **5** (1935/36), 228–260.
6. 'Some Remarks on "Facts" and Propositions', *Analysis* **2** (1935), 93–96.
7. 'Some Remarks on Empiricism', *Analysis* **3** (1936), 33–40.
8. (With P. Oppenheim) *Der Typusbegriff im Lichte der neuen Logik* (Wissenschaftstheoretische Untersuchungen zur Konstitutionsforschung und Psychologie), A. W. Sijthoff, Leiden, 1936.
9. (With P. Oppenheim) 'L'importance logique de la notion de type', *Actes du Congrès International de Philosophie Scientifique*, Paris, 1935, Vol. II, pp. 41–49. Hermann et Cie, Paris, 1936.
10. 'Eine rein topologische Form nichtaristotelischer Logik', *Erkenntnis* **6** (1937), 436–442.
11. 'A Purely Topological Form of Non-Aristotelian Logic', *The Journal of Symbolic Logic* **2** (1937), 97–112.
12. 'Le problème de la vérité', *Theoria* **3** (1937), 206–246.
13. 'Ein System verallgemeinerter Negationen', *Travaux du 19e Congrès International de Philosophie*, Vol. VI, Hermann et Cie, Paris, 1937, pp. 26–32.
14. 'On the Logical Form of Probability-Statements', *Erkenntnis* **7** (1938), 154–160.

15. 'Transfinite Concepts and Empiricism', *Synthese* **3** (1938), 9–12, Unity of Science Forum.
16. 'Supplementary Remarks on the Form of Probability Statements', *Erkenntnis* **7** (1939), 360–363.
17. 'Vagueness and Logic', *Philosophy of Science* **6** (1939), 163–180.
18. Articles 'Whole', 'Carnap', and 'Reichenbach', in *Dictionary of Philosophy* (ed. by D. Runes), Philosophical Library, New York, 1942.
19. 'The Function of General Laws in History', *The Journal of Philosophy* **39** (1942), 35–48.
20. 'A Purely Syntactical Definition of Confirmation', *The Journal of Symbolic Logic* **8** (1943), 122–143.
21. 'Studies in the Logic of Confirmation', *Mind* **54** (1945), 1–26 and 97–121.
22. 'Geometry and Empirical Science', *The American Mathematical Monthly* **52** (1945), 7–17.
23. Discussion of G. Devereux, 'The Logical Foundations of Culture and Personality Studies', *Transactions New York Academy of Sciences* **7** (1945), 128–130.
24. (With P. Oppenheim) 'A Definition of "Degree of Confirmation"', *Philosophy of Science* **12** (1945), 98–115.
25. 'On the Nature of Mathematical Truth', *The American Mathematical Monthly* **52** (1945), 543–556.
26. 'A Note on the Paradoxes of Confirmation', *Mind* **55** (1946), 79–82.
27. (With P. Oppenheim) 'Studies in the Logic of Explanation', *Philosophy of Science* **15** (1948), 135–175.
28. (With P. Oppenheim) 'Reply to David L. Miller's Comments', *Philosophy of Science* **15** (1948), 350–352.
29. 'Problems and Changes in the Empiricist Criterion of Meaning', *Revue Internationale de Philosophie* **11** (1950), 41–63.
30. 'A Note on Semantic Realism', *Philosophy of Science* **17** (1950), 169–173.
31. 'The Concept of Cognitive Significance: A Reconsideration', *Proceedings of the American Academy of Arts and Sciences* **80** (1951), 61–77.
32. 'General System Theory and the Unity of Science', *Human Biology* **23** (1951), 313–322.

33. *Fundamentals of Concept Formation in Empirical Science. International Encyclopedia of Unified Science*, Vol. II, No. 7, Chicago, University of Chicago Press, 1952.
34. 'Problems of Concept and Theory Formation in the Social Sciences', in *Science, Language, and Human Rights*. American Philosophical Association, Eastern Division, Vol. I. University of Pennsylvania Press, Philadelphia, 1952, pp. 65–86. (German translation, 'Typologische Methoden in den Sozialwissenschaften', in E. Topitsch (ed.), *Logik der Sozialwissenschaften*, Kiepenheuer und Witsch, Köln und Berlin, 4th ed., 1967.)
35. 'Reflections on Nelson Goodman's *The Structure of Appearance*', *Philosophical Review* **62** (1952), 108–116.
36. 'A Logical Appraisal of Operationism', *Scientific Monthly* **79** (1954), 215–220.
37. 'Meaning', *Encyclopedia Britannica*, 1956 edition, Vol. 15, p. 133.
38. 'Some Reflections on "The Case for Determinism"', in *Determinism and Freedom in the Age of Modern Science* (ed. by S. Hook), New York University Press, New York, 1958, pp. 157–163.
39. 'The Theoretician's Dilemma', in *Minnesota Studies in the Philosophy of Science* (ed. by H. Feigl, M. Scriven, and G. Maxwell), Vol. II, University of Minnesota Press, Minneapolis, 1958, pp. 37–98.
40. 'Empirical Statements and Falsifiability', *Philosophy* **33** (1958), 342–348.
41. 'The Logic of Functional Analysis', in *Symposium on Sociological Theory* (ed. by L. Gross), Row, Peterson & Co., Evanston, Ill., and White Plains, New York, 1959, pp. 271–307. (Italian translation published as a monograph: *La logica dell'analisi funzionale*, Istituto superiore di scienze sociali, Trento, 1967.)
42. 'Science and Human Values', in *Social Control in a Free Society* (ed. by R. E. Spiller), University of Pennsylvania Press, Philadelphia, 1960, pp. 39–64.
43. 'Inductive Inconsistencies', *Synthese* **12** (1960), 439–469. Also included in B. H. Kazemier and D. Vuysje (eds.), *Logic and Language*: Studies Dedicated to Professor Rudolf Carnap on the Occasion of his Seventieth Birthday, D. Reidel, Dordrecht, 1962.
44. 'Introduction to Problems of Taxonomy', in *Field Studies in the Mental Disorders* (ed. by J. Zubin), Grune and Stratton, New York,

1961, pp. 3–23. (Also: contributions to the discussion, on subsequent pages.)

45. *La formazione dei concetti e delle teorie nella scienza empirica*, Feltrinelli, Milano, 1961. (Contains items 33 and 39 of this bibliography, translated and with an introduction by Alberto Pasquinelli.)

46. 'Meaning', *Encyclopedia Americana*, 1961 edition, Vol. 18, pp. 478–479.

47. 'Deductive-Nomological *vs.* Statistical Explanation', in *Minnesota Studies in the Philosophy of Science* (ed. by H. Feigl and G. Maxwell), Vol. III, The University of Minnesota Press, Minneapolis, 1962, pp. 98–169. (Czech translation in K. Berka and L. Tondl (eds.), *Teorie modelů a modelování*, Nakladatelství Svoboda, Prague, 1967, pp. 95–172.)

48. 'Explanation in Science and in History', in *Frontiers of Science and Philosophy* (ed. by R. G. Colodny), University of Pittsburgh Press, Pittsburgh, 1962, pp. 9–33.

49. 'Rational Action', *Proceedings and Addresses of the American Philosophical Association*, Vol. 35, The Antioch Press, Yellow Springs, Ohio, 1962, pp. 5–23.

50. 'Carnap, Rudolf', *Colliers Encyclopedia*, 1962, Vol. 5, pp. 457–458.

51. 'Explanation and Prediction by Covering Laws', in *Philosophy of Science. The Delaware Seminar* (ed. by B. Baumrin), Vol. I (1961–1962), Interscience Publishers, New York, 1963, pp. 107–133.

52. 'Reasons and Covering Laws in Historical Explanation', in *Philosophy and History* (ed. by S. Hook), New York University Press, New York, 1963, pp. 143–163.

53. 'Implications of Carnap's Work for the Philosophy of Science', in *The Philosophy of Rudolf Carnap* (ed. by Paul A. Schilpp), Open Court, La Salle, Ill.; and Cambridge University Press, London, 1963, pp. 685–709.

54. *Aspects of Scientific Explanation and Other Essays in the Philosophy of Science*, The Free Press, New York; and Collier-Macmillan, Ltd., London, 1965.

55. 'Coherence and Morality', *The Journal of Philosophy* **62** (1965), 539–542.

56. 'Comments' (On G. Schlesinger's 'Instantiation and Confirmation') in *Boston Studies in the Philosophy of Science* (ed. by R. S. Cohen

and M. W. Wartofsky), Vol. II, Humanities Press, New York, 1965, pp. 19–24.

57. 'Recent Problems of Induction', in *Mind and Cosmos* (ed. by R. G. Colodny), University of Pittsburgh Press, Pittsburgh, 1966, pp. 112–134.
58. *Philosophy of Natural Science*, Prentice-Hall, Englewood Cliffs, 1966. Japanese translation, 1967; Italian and Polish translations, 1968; Swedish translation, 1969.
59. 'On Russell's Phenomenological Constructionism', *The Journal of Philosophy* **63** (1966), 668–670.
60. 'Scientific Explanation', in *Philosophy of Science Today* (ed. by S. Morgenbesser), Basic Books, New York, 1967, pp. 79–88.
61. 'Confirmation, Qualitative Aspects', in *The Encyclopedia of Philosophy*, Vol. II, The Macmillan Co. and The Free Press, New York, 1967, pp. 185–187.
62. 'The White Shoe: No Red Herring', *The British Journal for the Philosophy of Science* **18** (1967/68), 239–240.
63. 'Maximal Specificity and Lawlikeness in Probabilistic Explanation', *Philosophy of Science* **35** (1968), 116–133.
64. 'On a Claim by Skyrms Concerning Lawlikeness and Confirmation', *Philosophy of Science* **35** (1968), 274–278.
65. 'Logical Positivism and the Social Sciences', in *The Legacy of Logical Positivism* (ed. by P. Achinstein and S. F. Barker), The Johns Hopkins Press, Baltimore, 1969, pp. 163–194.

INDEX OF NAMES

SYNTHESE LIBRARY

Monographs on Epistemology, Logic, Methodology,
Philosophy of Science, Sociology of Science and of Knowledge, and on the
Mathematical Methods of Social and Behavioral Sciences

Editors:

‡JAAKKO HINTIKKA, *Models for Modalities. Selected Essays.* 1969, IX + 220 pp. Dfl. 34.—

‡PATRICK SUPPES, *Studies in the Methodology and Foundations of Science. Selected Papers from 1951 to 1969.* 1969, XII + 473 pp. Dfl. 70.—

‡D. DAVIDSON and J. HINTIKKA: (eds.), *Words and Objections: Essays on the Work of W. V. Quine.* 1969, VIII+366 pp. Dfl. 48.—

‡J. W. DAVIS, D. J. HOCKNEY, and W. K. WILSON (eds.), *Philosophical Logic.* 1969, VIII + 277 pp. Dfl. 45.—

‡ROBERT S. COHEN and MARX W. WARTOFSKY (eds.), *Boston Studies in the Philosophy of Science.* Volume V: *Proceedings of the Boston Colloquium for the Philosophy of Science 1966/1968.* 1969, VIII + 482 pp. Dfl. 58.—

‡ROBERT S. COHEN and MARX W. WARTOFSKY (eds.), *Boston Studies in the Philosophy of Science.* Volume IV: *Proceedings of the Boston Colloquium for the Philosophy of Science 1966/1968.* 1969, VIII + 537 pp. Dfl. 69.—

‡NICHOLAS RESCHER, *Topics in Philosophical Logic.* 1968, XIV + 347 pp. Dfl 62.—

‡GÜNTHER PATZIG, *Aristotle's Theory of the Syllogism. A Logical-Philological Study of Book A of the Prior Analytics.* 1968, XVII + 215 pp. Dfl. 45.—

‡C. D. BROAD, *Induction, Probability, and Causation. Selected Papers.* 1968, XI + 296 pp. Dfl. 48.—

‡ROBERT S. COHEN and MARX W. WARTOFSKY (eds.), *Boston Studies in the Philosophy of Science.* Volume III: *Proceedings of the Boston Colloquium for the Philosophy of Science 1964/1966.* 1967, XLIX + 489 pp. Dfl. 65.—

‡GUIDO KÜNG, *Ontology and the Logistic Analysis of Language. An Enquiry into the Contemporary Views on Universals.* 1967, XI + 210 pp. Dfl. 34.—

p.t.o.

*EVERT W. BETH and JEAN PIAGET, *Mathematical Epistemology and Psychology*. 1966, XXII + 326 pp. Dfl. 54.—

*EVERT W. BETH, *Mathematical Thought. An Introduction to the Philosophy of Mathematics*. 1965, XII + 208 pp. Dfl. 30.—

‡PAUL LORENZEN, *Formal Logic*. 1965, VIII + 123 pp. Dfl. 18.75

‡GEORGES GURVITCH, *The Spectrum of Social Time*. 1964, XXVI + 152 pp. Dfl. 20.—

‡A. A. ZINOV'EV, *Philosophical Problems of Many-Valued Logic*. 1963, XIV + 155 pp. Dfl. 23.—

‡MARX W. WARTOFSKY (ed.), *Boston Studies in the Philosophy of Science*. Volume I: *Proceedings of the Boston Colloquium for the Philosophy of Science, 1961–1962*. 1963, VIII + 212 pp. Dfl. 22.50

‡B. H. KAZEMIER and D. VUYSJE (eds.), *Logic and Language. Studies dedicated to Professor Rudolf Carnap on the Occasion of his Seventieth Birthday*. 1962, VI + 246 pp. Dfl. 24.50

*EVERT W. BETH, *Formal Methods. An Introduction to Symbolic Logic and to the Study of Effective Operations in Arithmetic and Logic*. 1962, XIV + 170 pp. Dfl. 23.50

*HANS FREUDENTHAL (ed.), *The Concept and the Role of the Model in Mathematics and Natural and Social Sciences. Proceedings of a Colloquium held at Utrecht, The Netherlands, January 1960*. 1961, VI + 194 pp. Dfl. 21.—

‡P. L. R. GUIRAUD, *Problèmes et méthodes de la statistique linguistique*. 1960, VI + 146 pp. Dfl. 15.75

* J. M. BOCHEŃSKI, *A Precis of Mathematical Logic*. 1959, X + 100 pp. Dfl. 15.75

Sole Distributors in the U.S.A. and Canada:

*GORDON & BREACH, INC., 150 Fifth Avenue, New York, N.Y. 10011
‡HUMANITIES PRESS, INC., 303 Park Avenue South, New York, N.Y. 10010